Grand Theories and Everyday Beliefs

Science, Philosophy, and Their Histories

Wallace Matson

OXFORD

UNIVERSITY PRESS

OXFORD
UNIVERSITY PRESS

Oxford University Press, Inc., publishes works that further
Oxford University's objective of excellence
in research, scholarship, and education.

Oxford   New York
Auckland   Cape Town   Dar es Salaam   Hong Kong   Karachi
Kuala Lumpur   Madrid   Melbourne   Mexico City   Nairobi
New Delhi   Shanghai   Taipei   Toronto

With offices in
Argentina   Austria   Brazil   Chile   Czech Republic   France   Greece
Guatemala   Hungary   Italy   Japan   Poland   Portugal   Singapore
South Korea   Switzerland   Thailand   Turkey   Ukraine   Vietnam

Published by Oxford University Press, Inc.
198 Madison Avenue, New York, New York 10016

www.oup.com

Oxford is a registered trademark of Oxford University Press

Library of Congress Cataloging-in-Publication Data
Matson, Wallace I.
Grand theories and everyday beliefs : science, philosophy, and their histories / Wallace Matson.
p.   cm.
ISBN 978-0-19-981269-1 (alk. paper)
1. Knowledge, Theory of.   2. Evolutionary psychology.   3. Belief and doubt.   4. Philosophy—History.
5. Science—History.   I. Title.   II. Title: Science, philosophy, and their histories.
BD161.M36   2011
190—dc22   2011004981

1 3 5 7 9 8 6 4 2

Printed in the United States of America
on acid-free paper

For George Bealer and Michael Della Rocca
sine quibus non

and in memory of
John Langshaw Austin

The whole of science is nothing more
than a refinement of everyday thinking.

Einstein, *Physics and Reality*

The facts of life do not penetrate
to the sphere in which our beliefs are cherished;
as it was not they that engendered those beliefs,
so they are powerless to destroy them;
they can aim at them continual blows
of contradiction and disproof
without weakening them.

Proust, *Swann's Way*

THANK YOU

To Amy Glick and the other Berkeley students who perked up when I tried out the high/low beliefs idea on them;

To the Rockefeller Foundation, at whose magnificent Villa Serbelloni much of the first draft was written;

To the UC Berkeley Committee on Research, which awarded me I've lost track of how many grants to keep on going;

To Randy Dipert, Hubert Dreyfus, Peter Hadreas, Stephen Hall, John Heil, Max Hocutt, Justine Hume, Candace Hyde-Wang, Adam Leite, Ruth Barcan Marcus, Rip and Rosary Matteson, Hugh Mellor, Gonzalo Munévar, Carlos Prado, Michael Reid, Larry Roberts, Walter Sinnott-Armstrong, Avrum Stroll, Bruce Vermazen, Albert Wald, Michael Williams, and Elizabeth Wolgast, for countless sensible suggestions and sensitive strokes;

To Drs. Joseph Young and Charles Turzan and the Kaiser Permanente HMO, who keep me alive;

To John Kekes, who was with me and guiding me at every twist and turn;

To Peter Ohlin of OUP for his gracious sponsorship, and appointment of the team Gunasekaran Hari Kumar and Jenny Wolkowicki to supervise a virtually painless production;

And to the friends at Yale who brought about the epiphany that transformed the final writing into a joyous labor of love.

CONTENTS

PART TWO: Miletus To Alexandria

PART THREE: The Legacy Of Christianity

Christianity, imposed by force on the ancient world, had a Grand
Theory based on a notion entirely foreign to Greek thought, that of
the Omnipotent Creator/Legislator (OCL). He could do anything
imaginable and was constrained by nothing. So there was no longer
anything that "could not be otherwise." Therefore, science in the
sense of proving necessity was impossible; it could at most only
investigate "second causes," what the OCL had hitherto allowed; and
this included miracles. Aquinas softened this doctrine somewhat by
dividing knowledge between Revelation (the soul and its fate) and
Reason (everything else), but the compromise was breaking down at
the beginning of the seventeenth century.

Descartes's attempt to shore up the Aquinian compromise and
the Real Distinction between soul and body by intellectual jujitsu
involving the hypothetical Evil Demon: a skeptical ploy going
beyond Pyrrhonism to cast doubt on all beliefs, low as well as high.
The Evil Demon's relation to the OCL. Reasons not to be scared
by the Evil Demon, his persistence in philosophy notwithstanding.

Hobbes's banishment of theology from philosophy and his
reassertion of the Milesian requirements.

Digression à la Hobbes on institutional facts, that is, beliefs
dependent on the existence of conventions. Rights. Sketch of a
consent theory of the State not dependent on an assumption of
equality among the consenters.

The Grand Theory in which God = Nature; ultimate reassertion of
the Milesian requirements.

Impossibility of maintaining the sovereignty of Reason on Cartesian
and Lockian assumptions. The "Problem of Induction": its
disappearance on rejection of the medieval doctrine of "logical
possibility."

23. Ethics Without Edification   *177*

  Bearing of the theory of high and low beliefs on ethics. Investigation of whether in fact a viable morality might exist without propping up by high beliefs.

24. L'Envoi   *197*

  Overview of conclusions reached in this book.

25. Conclusion?   *204*

  Reflections on whether the Milesian insight was a good idea for animals like us to have had, all things considered and *sub specie aeternitatis.*

Grand Theories and Everyday Beliefs

CHAPTER 1

☙

Introduction

How one thing led to another

Paul Grice said that at bottom there is only one philosophical problem, namely, all of them. Renford Bambrough said that philosophy is talk about what it is to be reasonable. These thoughts put together imply that the philosophical Big Bang is what Karl Popper named the Demarcation Problem: How to separate sense from nonsense.

The problem preoccupied the Logical Positivists of the first half of the twentieth century. A. J. Ayer notably held that only those propositions are meaningful that are verifiable in principle. "Purported" propositions, on the other hand, such that no possible experience could settle (or render more or less probable) whether they or their negations are true, are nonsense, to which terms of opprobrium like "pseudo-propositions" and "metaphysical" should be applied. They really have no meaning, cannot either describe or misdescribe reality, no matter how profound they may seem to be, or how many wars have been fought over them.

Though in sympathy with the aim of Ayer and the rest, I felt that they could not make their case stick, mainly on account of a Catch-22 embedded in that word "purported": plainly they had to know what a "purported" proposition meant before they could even begin to decide whether it meant anything. And it is just too implausible to claim, for instance, that "God's in his heaven, all's right with the world" has no *meaning*. Surely it is *true,* and consequently meaningful, if and only if God's in his heaven all's right with the world, just as "Snow is white" is true if and only if snow is white. But the Positivists had to deny meaning, not merely truth, to their rejections, because of their conception of philosophy itself, which is still current:

namely, that philosophy is the third member of the trinity of a priori disciplines (the others being mathematics and logic). Therefore the domain of philosophy, they held (or assumed), is limited to abstract unchanging things such as propositions—which are (in some sense) *there*, and either true or false and not both, eternally, whether anyone ever thinks of them or not—and verifiability-in-principle, a property that any proposition having it has eternally, whether or not anyone ever actually tries to verify it. Philosophy, then, could not on its own deny truth to a proposition—that function was the exclusive business of the natural, a posteriori, sciences.

The leading idea of the present book came when I asked myself, What if instead of propositions, we consider beliefs? And instead of verifiability, actual verification?

Such a study would have one advantage at least, over that of the Positivists: it would concern itself with indubitably real items in the real world (again, "assuming," if you like, that there is a real world). Nobody, not even Descartes, could doubt that beliefs exist, whatever else there may or may not be. Propositions, on the other hand, have to be argued for. Are they sentences? No—even though when one asks to be shown a proposition, what one gets is a sentence. Are they classes or sets of sentences? But introducing the notion of a class only multiplies the problems. Are propositions the *meanings* of sentences? That seems to be the favored answer; but just what is a meaning? What does it mean to ascribe existence to a meaning, a class, a set? There are philosophers with answers to all these questions, but the accounts are controversial and abstruse. Similarly with verifiability (and falsifiability) in principle. Some beliefs undoubtedly have in fact been shown to be true—at the very least, my belief that I exist. Verifiability-in-principle, however, is beset with problems. Verifiable by what? The senses, they say. But is it known for sure and a priori that the five (or six or *n*) senses of present-day terrestrial human beings are the only means of verifying anything? And so on.

For these reasons, and others that will emerge, it seemed to me that more progress on solving the Demarcation Problem was likely to be made by concentrating on beliefs and actual verification procedures. But is that really the problem the Logical Positivists were concerned with? Popper stated it not exactly in terms of distinguishing sense from nonsense, but of distinguishing science from pseudo-science. The Logical Positivists thought that this came to the same thing, since in their worldview only scientific statements had cognitive meaning. But that looks wrong. "God's in his heaven" is certainly not a statement belonging to any science (theology doesn't count); but also certainly, as we have observed, it means what it says, and that meaning is just as "cognitive" as "Snow is white" or "$e = mc^2$."

Well, if the Demarcation Problem truly contains all the "other" problems within itself, its statement should not presuppose any notions that could occur only at some stage of philosophical development that must come after the beginning, historically or logically or both. Here is an attempt: We want to formulate a rule that we can follow in allocating declarative sentences of the English (or any) language to one of two bins: the first, those describing beliefs that we have at least some reason to think correspond to the way things really are; the second, all the rest. This could of course be stated in other ways, for instance, what ought to count as evidence for truth of a belief, and the dichotomy would certainly be fuzzy, with lots of borderline cases, as in dividing people into bald and hairy, but worse.

The "historical plain method" of the physician-philosopher John Locke seems the proper way to go at such an inquiry: from early and simple cases up the scale of complexity. And since there could be no reason a priori to suppose that the development of knowledge was marked by discontinuities, it would be best to begin at the beginning: not with animals such as monkeys and dogs—let alone people—that clearly have beliefs, but lower down the great chain of being even unto sea anemones and jellyfish. And we would need to keep constantly in mind the question, What is belief *for*?

Anyway, it has seemed to me that this way of going at Epistemology, paralleling organic evolution, is worth trying out. If anyone should object that to do this is not to do *philosophy*, which by definition cannot be an empirical study, I would reply: It doesn't matter, grant or withhold any name as you please. Nevertheless, history is on my side. Spinoza was a philosopher if anyone ever was, and he had no qualms about proving propositions explaining (for example) racial prejudice. The fact is that what has been called philosophy has varied greatly from age to age. My own "definition" is such that Parmenides (not Thales or Socrates or Pythagoras, even though the last-named invented the word) was the first philosopher, Thales having been the first *scientist*; but my reason for doing so is convenience in exposition, nothing more profound. (And "scientist," by the bye, is a word not yet two hundred years old.)

I began literally at the beginning: with the (physical) Big Bang. (I *guess* that's the beginning.) I wanted to show how there came to be such things as beliefs, that is, at what rung of the evolutionary ladder creatures appeared that needed beliefs in order to further their *coping*—the central concept of evolution as I understand it. This turned out to involve speculations about the beginnings of language and the difference it made—in particular, the dichotomy of beliefs into two kinds, which I call *high* and *low*, the latter being those that *have been actually tested* by a procedure that I vaguely call "rubbing up against reality," the former those that have not been thus actually brought to judgment but are still up in the air; and how this *unnoticed* distinction works in human affairs. These topics occupy Part I, chapters 2–7.

Before I began my study I knew that what went on in Miletus at the beginning of the sixth century B.C. was the most tremendous revolution ever to occur in human thinking. Yet it surprised me to find how at that place and time three principles that jointly delineate the scientific outlook—monism, naturalism, and rationalism—had been present in the first non-mythological Grand Unified Theory of Everything, that of Thales. I then went through the history of Greek philosophy with a view to noting how these principles, and deviations from them, showed up and became institutionalized in the schools of Athens and Alexandria. This study occupies Part II, chapters 8–16.

The impact of Christianity on the Greek intellectual world was like that of an asteroid hitting the earth. Based on the idea of an Omnipotent Creator-Legislator, a conception not found in Greece even in pre-Thalesian times, it wiped out monism-naturalism-rationalism like dinosaurs. Creator-Creation were a fundamental dualism; the Creator negated Naturalism by running the world from outside; and most significant of all, the Creator being omnipotent and willful, there could be no more necessary truths about the world; everything became contingent at one fell swoop. The finding of what "couldn't be otherwise" could no longer be regarded as even a remote scientific ideal. Natural science, where it was tolerated at all, had to confine its efforts to the cautious study of "second causes."

But when the darkest Dark Ages were over and some works of Aristotle et al. were rediscovered, natural science experienced a sort of revival, and St. Thomas worked out a compromise dividing up the provinces of knowledge into Nature, where Aristotle was the Authority, and Grace—mainly the Soul—in which Revelation ruled supreme. On the whole this worked satisfactorily from the Church's standpoint through four centuries. But pressures built up and there was another crisis in the time of Galileo and Descartes. The latter's *Meditations on First Philosophy,* intended by their author to shore up the compromise, are traditionally thought to have inaugurated "modern" philosophy because the first two meditations present a skeptical argument, more influential than their intended refutation in the other four, that gave a *moderne* skeptical cast to subsequent thought: the mind-body problem, the external-world problem, and all that; and that in doing so philosophy reasserted its autonomy after having served its millennium-long stint as "handmaiden to theology." In my opinion, however, no such thing happened. The philosophy coming out of the Cartesian tradition left undisturbed two hangovers from the earliest Christian thought: the contingency of the world, and the interference in it *ab extra* of the Omnipotent Being—who, incidentally, served, slightly disguised, as the model for Descartes's bogeyman, who could have scared nobody if Omnipotence had not already become an unnoticed fixture of European thought, embedded in talk of "laws of nature" and "logical possibility." I show how this infection explains not only the odd metaphysics of Leibniz and Berkeley but even of so astute and untheological a thinker as David Hume. Thence the unhappy and unhealthy divergence of mainstream philosophy from science, which never abandoned the three Milesian requirements. But I supply a happy ending, sort of, when I point out that Hobbes and Spinoza both swam against the mainstream, preserving the Milesian tradition pure and indeed enhanced. These are the topics of Part III, chapters 17 –23. I have included also, as long digressions attached to accounts of Hobbes and Hume, two chapters discussing the beliefs about beliefs involved in institutional facts, and the theory of morality suggested, if not entailed, by the high/low belief distinction.

Chapter 24 is a reprise. Chapter 25, "Conclusion?" contains some ruminations on what I call the "final paradox:" whether the enterprise that Thales began was a good idea for animals like us to embark on, all things considered and looked at *sub specie aeternitatis.*

PART I

Before Miletus

CHAPTER 2

✧

A Brief History of Coping

L ife might be looked upon as an elegant solution to a certain problem: How to bring into being and preserve very complex *forms* over very long periods of time, out of the materials existing on the surface of the earth and in the conditions obtaining there.

To understand the implications of this viewpoint let us go back in thought to t_0, the Big Bang.

The world is full of a number of things. It wasn't then. There was only one thing, if conceivable as a thing: all the energy there ever was or will be, squeezed into one nanopinpoint, quite indistinguishable from nothing at all if *per impossibile* it had been perceived. The condition of zero entropy, that is, the most improbable distribution of energy. Then, the experts tell us, because of the instability of improbable states, there commenced the— from one perspective—downward slide to the more probable, but still astonishing, condition of the things, including us, now existing.

This has been a process of increasing differentiation and complexity, from zero difference, absolute simplicity (there was no room for *two* of anything) at t_0, to us, who coincidentally are the most complicated objects that we know about.

The first differentiation in the homogeneous expanding energy was into quarks, and the first complication—*structuring*—the buildup of these into the elementary particles through protons, neutrons, electrons, and neutrinos. With further expansion and cooling these formed into the simpler atoms, from hydrogen, No. 1, through (perhaps) iron, No. 26, many of which immediately combined with other atoms into molecules. Atoms more complex than iron could only be built up, it seems, in the rarely occurring circumstances of the explosions that we call novas, billions of years after t_0.

The uranium atom, 92 protons, 92 electrons, and (more or less) 146 neutrons, seems the most complex stable elementary structure that nature is capable of producing. The number of kinds of molecules that can be formed from atoms excluding carbon is large but not huge. However, the number of possible organic molecules—carbon compounds—is incalculably vast. Complete instructions for the construction of a living being are contained in a single molecule of DNA, of which there is a copy in nearly every cell of that organism, different from the DNA found in any other. But organic chemistry (past the first few steps) occurs only in very special conditions such as exist near the surface of the earth; nowhere else in the solar system as far as we know.

In a world with solids but no life, the most complex structures are crystals, from salt to geodes. If things do not merely lie around, as on our Moon, they engage in a mêlée of erupting, heating, cooling, washing, blowing, grinding, mashing, cracking, expanding, contracting, fusing, combining, decomposing, and exploding, as on Jupiter's moon Io. It would be misleading to call the process aimless, because that would suggest lack of something it might have had, whereas in fact, the notion of purpose would have no application whatever happened. Rocks stay in one piece if not hit too hard, they break up under stronger blows, and whichever happens does not matter.

Skipping over some further billions of years, we come here on earth to structures of carbon compounds that are *alive,* that is, have the power of self-duplication not merely into static copies as crystals do, but into beings with *metabolism,* to maintain themselves and to produce copies of themselves having similar capabilities. Some of the individual structures *cope*—develop their forms, starting from small and nonviable beginnings, to individual viability (growth); incorporate materials from their surroundings into themselves for growth and replenishment of energy (nourishment); preserve their forms intact by avoidance of and active resistance to forces tending to break them down (defense); and create from themselves other individuals of the same form, capable of continuing in their life spans the same processes (reproduction). Accomplishment of these functions is *success,* frustration is *failure;* what aids coping is *good,* what interferes with it is *evil;* the activities of coping are what the organism *ought* to do.

The italicized terms above are, of course, relative to the particular organisms: what is good for the big fish (swallowing) is evil for the little fish (being swallowed). Nevertheless, these terms are objective in their application. It is part of the literally true description of the world containing organisms that some are succeeding, some are failing; these facts are independent of how the organisms feel, or indeed of whether they feel at all. On the other hand, it is only by metaphor that we speak of a volcano as "failing" to erupt.

There is nothing that a volcano *ought* to do. In other words, it is life that introduces *value* into the world, and *doing* as opposed to mere happening.

If organisms succeed repeatedly in coping, their forms may last indefinitely even though every particular exemplification of the form is ephemeral. Though archy the cockroach is but a brief candle, The Cockroach has been around for a time longer than the Pyramids and the Alps by many orders of magnitude.

However, the random elements introduced into reproduction by sex, cosmic rays, and whatever, ensure that the forms of offspring will not be exact copies of parents but will vary somewhat, and in some ways that favor or disfavor coping. Consequently, some will be more successful at reproducing than others; and given that descendants are closer in form to parents than to others of the same genus, the favorable variants will predominate in succeeding generations. This is the *survival of the fittest.*

This phrase has been objected to as a tautology. Since comparative fitness, it is argued, can be measured only by survival rates, it reduces to "survival of the survivors." But this misinterprets the slogan. The survival in question is that of a species, while "fitness" refers to the individual's ability to cope. It could be, and sometimes is, the case that the fittest individuals of a population have a comparatively low fertility, for various reasons, so that the stock as a whole degenerates, making an exception to the generalization. Moreover, Darwinian fitness is implicitly limited to conditions of free competition in a stable environment. In their day the "fittest" animals were dinosaurs. They (the big ones anyway) did not survive the asteroid collision of 65,000,000 B.C., but that was not their fault.

By *form* of an organism is to be understood not merely its physical shape and other characteristics but also its typical *behavior,* using this word in the broad sense of all its internally generated motions. There are two ways in which organic form can be modified, one fast and one slow. The slow mode is by natural selection or, when man intervenes and speeds it up a bit, breeding. These can affect both shape and behavior. The fast mode, affecting only behavior, is *learning*: the individual organism concentrates on the modes of behavior that have proved successful in coping and discards the detrimental ones. Although the enhanced adaptive capabilities resulting from learning cannot be passed on genetically, an ability to learn can be selected for.

HOW PLANTS COPE

The coping of plants is entirely passive. Literally rooted to the spot, they can only hope, so to speak, that adequate light, water, nutrients, and genetic materials will come their way. They cannot learn; they can undergo adaptive modification only by the slow way of selection. Nevertheless, some plants *do* something: there is such a thing as plant behavior, for example, the closing of blossoms at nightfall. Plants receive information (as distinguished from mere tweaks and blows) from the outside world and process it. The ingenuity of the devices that nature hath wrought in this manner is marvelous, climaxing in the Venus flytrap's trick of meshing its deeply serrated leaves almost instantly, thus capturing an insect whose nitrogen it can then absorb, making up for nitrogen deficiency in the soil. However, even this spectacular vegetable accomplishment is a *re*action to an environmental factor rather than an action by the plant. There is a direct linkage between an external occurrence, the insect's crawling over the "hair trigger," and the accelerated differential growth that closes the leaves. Any (repeated) stimulation of the trigger will be followed by the reaction; the plant does not distinguish between nutritive objects and useless or harmful stimuli on the other hand; it relies on the preponderance of the former in its environment. Plant tropisms are analogous to animal reflexes: direct and stereotyped responses to stimuli, without the mediation of the central nervous system, though their mechanisms are very different.

Other stratagems for survival in plants have developed—for instance, the dandelion's rapid growth of broad leaves touching the ground at the base of its stem, which effectively prevents rival plants from getting established in its vicinity and competing for soil nutrients. But such accomplishments cannot be counted as things the plants *do*, even in the broad and relaxed sense of behavior in which we count tropisms. Modern dandelions zap their competitors in this relentless manner because evolution has favored those individuals that were "better" at it, starting from proto-dandelions that only occasionally exhibited a slight propensity for this kind of growth in a few individual specimens. Production of numerous seeds, development of characteristics enticing insects to do their part in fertilization, various mechanisms for promoting widespread distribution of seeds (so that the chances of some being implanted in suitable soils are increased) are other such evolutionarily furthered adaptations.

Adaptation is passive; nothing the organism *does* contributes to the process, which depends on individual variations in DNA. It may be driven internally, as when the broad-leaved dandelions supersede the others; or externally, as when the onset of an ice age bestows an advantage on trees with needle-like leaves. Adaptation is not learning. But as it is the analogue in the species to learning in the individual, it may be regarded, metaphorically, as a learning process in which the genotype is the pupil. A species is presented with a problem recurring in each generation; each succeeding generation copes with it more efficiently than the previous one did. All living creatures have "learned" in this way; it is the only way available to plants. Necessarily it is a very slow process, unsuited to dealing with crises and rapidly changing conditions. That is the main reason why 80 percent or even 90 percent of all the species of organisms there ever have been are now extinct.

It would be a mistake to suppose that adaptation is a process in which organisms start out poorly adapted and end up well adapted. Adaptation is a relative term: an organism is well adapted when it functions efficiently in the environment in which it lives. Normally, that is, in the absence of cataclysmic environmental change, all organisms are well adapted.

ANIMALS

The originally defining characteristic of animals, and their principal advantage over plants, is their ability to move from one place to another in search of nourishment and mates and in fending off or avoiding predators and other dangers. This motility is the primary kind of organic *activity*; it is the source of consciousness, subjectivity, learning, and of *belief,* as we shall see.

To be of use for coping, motility must be linked to some means of distinguishing features of the environment. Think of a simple aquatic animal that can survive in fresh water but not in salt. Motility will be a *dis*advantage, indeed a lethal one, if, happening to live in an estuary where fresh river water flows into the salty ocean, the organism cannot distinguish between fresh and salt. Hence to be viable it must possess some *internal indicator* of the difference, and there must be some linkage between this indicator and whatever in its physiology controls the direction of its swimming: call it the animal's "rudder."

It would suffice, in this case, to have a direct chemical feedback connection between indicator and rudder: an organ, activated by sodium ions, which would position the rudder in the direction of decreasing sodium ion concentration. Such a device would be like the photo-tropism of the sunflower. All animals are endowed with some such tropisms or *reflexes*, and very simple animals have no other ways of guidance in getting about. There need be no conscious awareness in the animal, for the organism makes no decision; it automatically behaves in the manner increasing the likelihood of survival. This simplest and most direct response, the reflex arc, is a definite, virtually unmodifiable, and rapid reaction to a certain type of stimulus, bypassing the higher brain centers (if there are any).

Some simple animals such as coelenterates behave only reflexively, making the same response to whatever disturbs them: closing in the sea anemone, stinging in the jellyfish. More complicated creatures, ourselves included, are equipped with some reflexes of this type: the knee-jerk and the adjustment of the iris to intensity of light, for example. They are hard-wired programs for coping directly with certain types of stimuli. When the patella is struck the knee *always* jerks (if the nervous system is functioning properly—hence the use of the reflex in diagnosis). The jerk cannot be suppressed no matter how hard one "tries," and it is elicited by no stimulus other than a blow to the patella. Other reflexes such as salivation and erection, which are mediated by sense perception, will be discussed later.

INSTINCT

Possession of distance receptors enables an animal to pass beyond reflex behavior and make more efficient use of motility. Animals equipped with one or more of the distance receptors, sight, hearing, and smell, and with musculature to steer themselves, can detect food, mates, and dangers that they are not in contact with, and can move appropriately toward and from-ward. Detecting is discriminating, as in picking out the edible bits of the environment from the inedible background; the members of one's own species of the right sex, age, and estral condition for mating from the non-qualifying animals; and the dangerous from the harmless. Making these discriminations consists in receiving and processing information from the world: *sensation* and *perception*.

From the sea anemone and jellyfish, with a few nerves activated by pressure and specific chemicals (so that they may be said to have rudimentary senses of touch and smell), to the moth, pursuing the female moth and fleeing the bat, is an enormous complication and refinement of sensory equipment, paralleled by the intricacy of the moth's resources for response. As its picture—dangerous metaphor—of the world, the anemone's internal indicator is perhaps painted in but two or three strokes, corresponding to pressures on its integument, whereas the moth's is a detailed mosaic in which we, could we but experience it, might recognize similarities to our own sensory manifold. Both, however, are maps adequate to orient the kind of effort the animal is capable of exerting.

Sufficient information for coping may consist merely in receipt of a certain signal, if the animal is genetically hardwired to react to that indicator in an appropriate manner. If, for instance, among the inputs to the male moth's organ of smell a certain pheromone is present, he will fly up the gradient of increasing intensity until he encounters the female of his species and mates. This behavior is *instinctive*.

Instinctive behavior is like reflex in following straightway and automatically upon the occurrence of a specific stimulus, and in being unlearned, difficult if not impossible to modify, and manifested by every normal member of the species; but it is different in several respects. Reflex motions are mainly defensive and rapid, virtually instantaneous; instinctive behavior, the greater part of which is concerned with doing things needful for carrying on the species, may spread almost indefinitely over time: web spinning, nesting, migrating. A reflex is a motion of some particular part of the body; instinct involves the body as a whole. Instinctive behavior is subject to inhibition: people, and some beasts, can be trained to refrain from carrying through some instinctive patterns.

Instincts are results of the slow "genetic learning" of which we spoke earlier. And so, since they are solutions to problems, it is tempting to think of them in terms of their "goals": birds fly south in winter *so that* they won't starve and freeze; *the purpose* of the beaver dam is to form a lake. But an animal—including the human—engaged in instinctive behavior is doing it with the *feeling* that he or she *has to*, without reference to outcome. One *finds* oneself behaving in that way, and if prevented, one feels uneasy until the obstacle is removed. Sex, viewed from outside, is for propagating the species, but desire to achieve this end is only sometimes why people engage in sexual activity. This point is important for our inquiry, as it indicates that instinctive behavior does not involve expectation: that is, inasmuch as expectations are future-oriented beliefs, it is not a necessary condition for instinctive behavior that the animal should have any beliefs about the consequences of what it is doing.

The sensing that triggers an instinctive response is (in us at least) conscious, and often complicated, requiring discrimination of a particular stimulus from among others in the same sense modality: "filtering," as it were. Mating is instinctive, but the male must be able to separate out the female of his species from other creatures of sometimes quite similar shape, color, or odor. And this ability to discriminate is innate, not learned; like the program of the behavior itself, it is hardwired.

Sensing is the acquiring of information from the world through the world's effects on the organs of sense. Sound, rapid undulant motion of the air, impinging on the eardrum, sets that membrane vibrating in the same frequency. Through the Rube Goldbergian linkages of the inner ear, the aural nerve is caused to transmit electrical vibrations of corresponding frequency into the brain, where events occur that are experienced in the mode of hearing. This is the sensing part: the activation of an internal indicator.

What happens next depends on the nature of the sound and of the animal hearing it. If the former is bat sonar and the latter is a moth, the moth will forthwith engage in evasive maneuvers. She will do so whether or not she has ever heard a bat before, and whether or not she has previously flown in this fashion, so neither the behavior nor its stimulus is learned; it is an instance of hardwiring, that is, instinct. An internal indicator has triggered the evasion response. The word "indicator" must not seduce us into supposing that something is being read by a mothunculus. We cannot imagine what is going on in the mothly subjectivity (if there is one). All we know is that the moth is responding appropriately to a feature of the world that is registering within her.

The same might be said of an artifact—for instance, a guided missile programmed to take evasive action when an onboard receiver indicates illumination by enemy radar. Is there any structural difference between the two processes? None is readily apparent. In both cases the stimulus/behavior linkage is innate, built-in, hardwired, comes with the package. In the one case, evolution selected the moths that were good at evading bats. That quality consisted in quick recognition of bat-in-vicinity, coupled to immediate evasive maneuvers. Analogously, the engineers of Martin Marietta built a radar sensor into the missile and linked it to a programmed series of course changes. The missile does no thinking, neither does the moth. Nor does the missile feel anything (or so we believe). What about the moth? As far as the missile analogy goes, there seems no reason to suppose that the moth is in any way aware of what she is doing, that there is any mothly subjectivity.

But there *must* be. If stuck, cut, or hit, the moth will writhe and flutter her wings. This is *pain* behavior. Only a Descartes could resist the inference that the moth *feels* something—and something unpleasant, too. There is no counterpart to this in the missile.

Every motile animal when damaged moves in a way appropriate for protecting itself as well as it can. We know from our experience that pain is the stimulus for these maneuvers when *we* engage in them, and nothing but pain—unpleasant sensation—could serve in this role, because only pain has a built-in urgency that keeps it from being overruled unless for extremely important reasons. So the world *hurts* moths, and other creatures down to earthworms and no doubt farther. This should not surprise us. Motile animals obviously could not cope without strong avoidance signals.

To this extent, then, it is not fanciful to infer the existence of mothly subjectivity: there is a "what it is like to be a moth," which we cannot imagine, but which must include unpleasant (at any rate to moths) feelings, and (perhaps) the feelings associated with the internal indicators we have postulated. But no memories, no expectations.

Perhaps, as Anaxagoras speculated nearly 2,500 years ago, pain is the original and basic feeling from which all the senses have evolved. To this day, overloads in any sense modality hurt. Vision developed from a light-sensitive skin spot; but what would the evolutionary advantage of that original spot have been? It might have been that light *felt painful*, so that the possessor of the spot tended to retreat from regions of light to dark, where for one reason or another conditions for survival were more favorable. Why not "felt pleasant, stimulating the organism to move into the light"? Well, that is not the way the world works; see Schopenhauer. Pleasant sensations, as distinct from states or conditions, are comparatively rare, and never are *stimuli*—consider the etymology of the word. *Anticipations* of pleasure motivate behavior; but capacity for anticipation is to be found only among the more advanced animals.

To return to the moth, which we are letting stand for any animal all of whose behavior is either reflexive or instinctive: moths have sense organs, feeding them information which—since they possess a subjectivity—they *feel*, in a broadened sense of "feel" in which we might say that green is the way the eye feels grass. When a moth sees a tree, the visual pattern of her internal indicator may be as complex as ours; but whereas we can *contemplate* a tree, perhaps all that registers in the moth is a feeling such that if we tried to put it into words, we might come up with something like "I must lay my eggs on the underside of that leaf." The consciousness (such as it is) of an instinctive creature must be, so to speak, a categorical imperative: imperative because it is entirely bound up with practice, with what is to be done; and categorical because, as we have observed, instinctive acts are not done in anticipation of their consequences. Persons acting under the influence of post-hypnotic suggestion, when asked why they are standing on their heads, tend to reply that "they just feel they have to."

On the irritability theory of original sensation, the basic subjective rendering of the world is not so much of how it is as of how I ought to change it—such as "This spot is too hot, I must get away from it." Of course, the creature has no concept of "I," nor anticipation of how it might feel elsewhere. It just experiences an urge-to-cool-off feeling, or something even simpler. Moths cannot enjoy the luxury of disinterested contemplation from a vantage point on the fact side of the (alleged) fact/value gap. Nevertheless "It's hot" is a component of the feeling, and to that extent the moth is aware of how the world *is* at that place and time. Indeed, possessing, as she does, complex and efficient eyes, a detailed representation of her environs may be subjectively present. Or consider the spider about to spin a web: the inner representation of her surroundings must be sufficiently detailed for her to discern points of possible anchorage.

CHOICE

Animals behaving only reflexively and instinctively have no need of anticipation, as we have seen, nor of memory. Their living is strictly in the present, with nothing in them corresponding to any notion of time. Or hardly anything. If they are to distinguish moving from stationary objects, they must to that extent be capable of differentiating a before and after in time, which means that the sensation must "die out" not instantaneously but with a certain gradualness, as on a radar screen. In this lies the germ of memory, the notion of a temporal sequence.

The difference between creatures with memory, and those (almost) lacking it, is hardly less profound than that between living and dead. Memory makes possible action (as opposed to reaction), choice, learning, goal-seeking, anticipation, perception (as opposed to mere sensation), hope, fear, intention (and intentionality), meaning, and belief. All these notions are interconnected. Let us begin our discussion with learning.

What happens when a rat learns a maze is something like this: The hungry animal positioned at START makes a sequence of motions, initially random, terminating by chance in access to the supply of food at EXIT. When later the animal, hungry again, is placed back at its original starting position, it moves less erratically to where the food is. After a further number of repetitions the rat goes directly to the food, making no wrong turns at all. It has come to remember which turns are right and which wrong, and to avoid the latter.

This is a great leap upward from moths. Maze running is goal-directed behavior, which reflex and instinct are not. The rat makes these motions *in order to* get the food. Because it can anticipate a goal in imagination and remember the sequence that in past brought it to that goal, it can learn to cope with a novel situation.

A maze-running instinct could in principle be achieved by natural selection, if the world were full of identical mazes with rat food available nowhere but at their exits—but only at the cost of untold millions of starved rats that were not yet hardwired to make the correct turns fast enough; and the race of survivors, who ran the maze correctly, would perish if the maze design were altered, their hard-won instinct being of no use in the new conditions.

Learning involves both memory of the past and anticipation of the future, but asymmetrically. Dogs' and cats' memories may go back for many years: witness Odysseus's dog Argos after twenty years wagging his tail upon his master's return, then dying—a feat with less spectacular but still impressive analogues in the experience of every pet owner. Their future, however, is very constrained—more or less literally to the next meal, the imagined goal of their present behavior. Dogs and cats take no thought for the morrow, much less for next week or year. The long-term provisions that birds and beavers make for what is to come are instinctive, not planned. Learning also involves self-direction of behavior: the making of *choices*, resulting in the performance of *actions* in the narrower sense in which they are distinguished from the automatism of instinct, to which our attention has been limited up to this point.

There has been an evolutionary progression or complication from

- animals that can react only to direct gross pressure on their integuments (direct reflex), to
- animals with specialized organs inputting information originating at a distance, some of which information is filtered by built-in templates in the nervous system from other similar information not bearing on the animals' vital interest, which filtered information triggers automatic, localized responses tending to preserve the animal and/or the species to which it belongs (sense-initiated reflex), to
- animals in which certain filtered informational inputs produce feelings in response to which innately programmed behavior, furthering the interest of the animal and/or its species is elicited (instinct), to
- animals capable of evaluating input information and initiating action which, they anticipate, will better serve their interests than other courses that they might have initiated (choice).

One can see in a general way how the development must have gone from step (i) to step (ii) to step (iii)—direct reflex to differentiated reflex to instinct. But what about the transition from instinct to choice? Presumably, refinement of sense organs, with improvement in memory, so that comparisons of outcomes of different courses of action that had been engaged in would thrust themselves upon the animal, biasing it to act next time according to the pattern that had proved more advantageous on the previous occasion: this would be the genesis of *reinforcement* and of learning. At first, all actions would still be instinctive, but some would be, as it were, more instinctive than others, and their patterns would prevail in cases of conflict. They would come to be favored on account of their felicitous outcomes and adopted in preference to alternatives.

A reason why merely instinctive animals cannot be trained is that training involves reinforcement. A correct response is reinforced by arranging things so that the animal satisfies a *desire* by the response. But desires involve anticipation, and because instinctive animals have no anticipation, they have no desires, even though they have wants and needs, which would be desires if only the animals were capable of anticipating their satisfaction and of envisaging that satisfaction as connected to their behavior.

The gap between instinct and choice is further narrowed if we consider that not all instincts—and in higher animals not the most important ones— are hardwired programs to behave, straight off, in a particular way, but are programs to *learn*, at a certain stage of development, a certain kind of behavior. Notoriously, birds have to learn to fly; at least they are not awfully good at it first time out of the nest. What is innate in the chick is the urge to try it, and the aptitude to get it right after a few tries. Similarly, we are innately programmed to learn to speak a language, though what language is fortuitous. Still, flying and speaking are properly said to be instinctive in the animals that do these things. They are not chosen behaviors, even though elements of choice enter into their development. Perhaps they should be regarded as instinctive behaviors, the experiential cues (triggers) for which are particularly complicated and extending over long stretches of time. In any case, they represent a transitional stage between simple instincts (suckling, exhibited at birth) and behavior with no innate component (voting in a presidential election).

It is in the transition from (iii) to (iv) that *imagination* first enters upon the scene as, so to speak, memory of the future. The animal remembering the outcome of past action A, and of B, and comparing them, is in that very act imagining what it *would be like if* it performed A again or B again. This imagined scenario is the animal's *anticipation* of the consequences of A and B, which is—Eureka!—its *belief* as to what states of affairs A and B will bring about. If, all things considered, the anticipated consequences of A are more *desired*—imagined with a more positive feeling tone—than those of B, and the alternatives are exclusive and exhaustive, the animal *chooses* A and performs it, straightway or as soon as circumstances permit.

BELIEF

Let us pause to catch our breath. We have not discovered what belief is, in general terms, but we have come upon one kind of mental condition that plausibly falls under the concept, namely, anticipation or expectation, and we have roughly located the step in the zoological ladder on which it may first be discerned. As we have had more than one occasion to note, animals whose behavior is only reflexive or instinctive need not anticipate the out-comes of their doings, nor indeed have any notion of the future, since what they do is determined by their innate, hard-hardwired or "protected" programming. Only animals capable of altering their own programs will have any use for such notions. For them, however, these conceptions will be not merely useful but indispensable.

A NOTE ON TRUTH

It is in accordance with standard English usage to say of expectations, and of beliefs in general, that they are true or false (or in between). And since dogs and cats, which have no language, nevertheless have true and false expectations, truth and falsity cannot be merely linguistic.

My theory of truth, such as it is, might be called a "genetic correspondence" theory. While sentences, propositions (whatever they are), judgments, and so on may be truly said to be true or false, they are so derivatively from this basic application to the relation between a belief and what it is about. A practical belief is a brain configuration (conscious or unconscious, occurrent or standing) that has evolved (or the tendency to form it has evolved) as an aid to coping. It may or may not be structurally similar (as for example the magnetization pattern of an audio audiotape is similar to the Moonlight Sonata) to what it is about. If it is, it is true; if not, false.

But this truth admits of degree. The dynamic situation "out there" and its indicator or representation "in here" will at best be in imperfect correspondence. The "law of excluded middle" does not come into play at this stage.

The basic test for truth is acting, in the process of coping, as if the belief is true. If the coping goes OK, *when it is action in the context of what the belief is about,* that is positive evidence for its truth (again, more or less).

With anticipation come hope and fear, love and hate, and in due course the more complex emotions such as greed, envy, benevolence, and gratitude. Instinctive animals know none of these. The famous wasp that lays her eggs beside bugs that she has paralyzed with her sting is providing for the nourishment of her offspring, but not out of maternal affection; it is simply instinct. Animals do not make reflex motions and engage in instinctive behavior in order to achieve end states imagined in advance. Choices, by contrast, are always so motivated.

However, ability to choose feeds back in to the nature of reflex and instinct. The *conditioned reflex* is such a modification. In simple animals the necessary and sufficient external condition, the "trigger" for eliciting reflex behavior, is as unalterable as the behavior itself. But in animals capable of learning, the triggering function may be assigned to substitute signals virtually ad lib.

The salivating of the hungry dog when food appears is a true reflex, elicited directly and uncontrollably by the stimulus. Professor Pavlov trains him to salivate upon hearing a bell, even when no food is yet in the offing. The behavior is uncontrollable at every stage in the training: there is no element of choice in it from beginning to end. Yet learning is going on. How can this fact be reconciled with the alleged immutability of reflex?

What the reflex link is *between* is internal indication of food (not food *simpliciter*) and salivation. In the merely instinctive animal the internal indicator of food is the sensing of food (or of what is close enough to food in appearance to fool the animal) and of nothing else. In the dog, however, an animal with memory and the capability to anticipate, perception of things other than food can indicate that food is in the offing—can make the animal "think of" food, as we say; and this is sufficient to trigger the salivation reflex. (Reader: At this point, imagine that you are about to be forced to drink a glass of undiluted and unsweetened lemon juice.) For his experience may have produced the expectation that ringing of the bell will be followed right away by dinner. "Internal indicator," "belief," "expectation," and such-like "mentalistic" terms are to be understood (in causal contexts) as signifying postulated neural structures and events. That anticipation is involved, is shown by the possibility of extinguishing the reflex by ringing the bell and *not* providing food: after a certain number of disappointments, the dog ceases to slobber any more when the bell rings.

Learning has an even more profound effect on instinctive behavior than on reflex: it can, in many cases, suppress it. Instincts themselves are not, strictly speaking, behavior, but behavioral biases and tendencies; many things can interfere with their manifestation. Chrysippus the Stoic philosopher trained hunting dogs to ignore hares. As St. Augustine noted, men cannot be trained not to have erections (which are reflexes), but through education they may come to refrain from fucking (instinctive behavior) and even to abhor the very notion.

For various reasons, which we shall note later, it is easier to suppress instinctive behavior in *Homo sapiens* than in any other animal. This led some psychologists to deny that there are any human instincts at all save suckling and grabbing-when-falling. There is no such thing as human nature, they held, if by that is meant innate and ineradicable tendencies to behave in definite ways. People are entirely plastic. Why this view is mistaken, will be discussed in chapter 23. At this point we note only what is well known to all animal trainers, that the inhibition of instinctive behavior can never be counted on to be permanent: the counter-biases introduced in the training must be continually reinforced lest the animal "revert to nature."

Like the instinct to learn—which by definition only an animal can have whose behavior is not all merely instinctive—is instinctive evaluation. As merely instinctive animals have no desires, neither do they have any preferences (save the negative preference for pain). However, instinctive preferences, in choice of mates, for example, are rife among vertebrates. There is even some evidence that it is instinctive in *Homo sapiens* to prefer, ceteris paribus, to live in open country within sight of a body of water.

RECOGNITION

Sense perception itself is radically transformed by possession of memory. Animals such as spiders and flies, which cope by reflex and instinct alone, may have quite complex and sensitive sense organs, which, however, function principally as on-off devices for instinct triggers. When certain information passes, the resulting internal indication activates a program of instinctive behavior. Other data are discarded; there is no data storage. The organism, then, can make no comparison of input save with its built-in instinct "templates," and the hundredth time the reaction is elicited is no different from the first. A moth might be said to be cognizant of the fact that a bat is pursuing her; the bat sonar is heard, triggering the evasion instinct, but the moth does not "re–cognize" the bat.

Contrast a case in which the sound is made by a growling tiger and the hearer is a boy, who flees. Just as with the moth, there is an internal indicator; but unlike the moth, the boy does not flee forthwith and automatically. He is not hardwired to behave in a definite way upon hearing sound of this timbre; that is to say, there is no instinct or reflex. He flees because he re-cognizes the sound as meaning "Tiger!," and, moreover, as meaning danger in this particular context; he would not flee if he heard the sound in a zoo. He has learned to make this connection, either by having heard tigers growl and having stored some semblance of this experience in a memory trace now activated by the sound; or perhaps he has only heard his father imitate the sound, at the same time directing him to run away if ever he hears it outdoors. He *imagines* an approaching tiger which will pounce on him and devour him unless he quickly gets up a tree: that is to say, he *fears* the tiger, and fearing is expecting, and expecting is believing. That is why we can say that the boy *acts*—performs an internally initiated motion—in the *belief* that he is threatened by a tiger. These elements are absent in the moth and missile cases, which for that reason are instances of instinct and cybernetics, respectively, not involving belief.

Boys living in tiger-infested country, if they survive, will reach the point at which the hearing of the growl will be followed forthwith and automatically by flight, "without thinking." They will have acquired what amounts to an instinct manufactured ad hoc which dispenses with the necessity of having to rethink a routine every time the occasion for it arises. This is the point of all learning-how, all acquisition of skill.

Skills, however, are under the control of the performer, who *can* stop or modify the performance at any point. And they are performed for a purpose recognized by the performer. These features distinguish skills from the other type of learned response, the conditioned reflex.

All motile animals are sentient, that is, they receive information from the environment and modify their motions accordingly. It does not follow, however, that they are *conscious*—have an inner life or subjectivity. As we know from our own experience, reflex behavior is not conscious. Instinctive behavior sometimes requires sizing up, and sizing up requires memory, but it does not have to be conscious memory. Past experience influences the manner in which we presently size up a situation, but it is not necessary that when we do this, we form a mental image of what occurred on some past occasion. Only when behavior involves anticipation of the future does consciousness become indispensable. Being able to anticipate the future, and being able—selectively and for a reason—to delay reaction to a present stimulus, amount to the same thing; this ability is the genesis of consciousness.

Spiders don't need to be conscious, if all their behavior is instinctive, consisting in responses to stimuli without any anticipation. If that is so, then although their actions have objects, for example, capturing the fly and then ingesting its bodily fluids, it cannot be said that these actions are *about* anything; the spider has no intentions. *We* say the web is intended for capturing insects, but the spider just spins it. Aboutness or intentionality, like consciousness, is a matter of anticipation. The germ of aboutness lies in the creature's relation to the not-yet-existent state of affairs to which its premeditated action is to apply. Deciding to do this rather than that requires the internal rehearsal of actions not yet performed, and comparison of their foreseen outcomes. The chooser must somehow produce internal simulacra, which, following the tradition and for lack of a better term, we shall refer to as "mental images." (A better metaphor than "mental image" would be "mental stage set." One can enter a set, unlike a picture, and act in it. This feature might help to undermine the passive-spectator view of knowledge endemic in philosophy. We shall, however, continue to speak of images, and the faculty of producing them as "imagination.") They are not, of course, fabricated ex nihilo, but constituted from materials in the memory. Yet expectation, anticipation, and rehearsal of action are entirely different from remembering. They open up another dimension to experience: the future. There is no reason why an animal incapable of choice should hope for anything or fear anything. Such a creature's sentience would have the sole function of triggering instinctive reactions.

Cats and dogs make choices; consequently, they have inner lives, with emotions and the ability to imagine actions not yet performed and their probable outcomes in conditions toward which the present setup is perceived as tending. A hungry dog turning from a bowl of food to obey its master's command to run and catch the Frisbee that he is about to throw, exhibits all these elements. The dog leaps toward the point where the Frisbee will be, not where it is. They add up to beliefs about what is going to happen—a conception of the future. However, it is a shallow conception; their "future" can hardly extend appreciably beyond the specious present. Migrating, nest building, web spinning, and so on, are activities with reference to a more remote future, but they are instinctive.

Projecting memories to form expectations—"induction"—can be epistemologically hazardous. However, the risk tends to be minimized by this shallowness of beastly imagination. Canine judgment as to what is likely to happen does not have to wait long to be put to the test, to be rubbed up against reality. A puppy believing that a porcupine will enjoy being romped with, learns of its error straightway. A tendency to form erroneous beliefs is a biological disadvantage, if the beliefs in question are such that choices based on them expose the believer to peril. Evolutionary pressures will therefore weed out from the animal stock individuals that are burdened with such a tendency, with the result that dangerously false beliefs will seldom occur.

The content of a belief of the simple kind we are now discussing is an internal schema of the way things are and are going to be. It is *true* if things really are that way, *false* otherwise. It is put to the test, rubbed up against reality, jeopardized, when the animal holding the belief acts in a way that is appropriate (that is, furthers the animal's interest) just in case the belief is true. If the belief is false, the action will be frustrated, the animal will experience surprise at least, injury or death at most; and in any case it will cease to hold the belief (with qualifications and exceptions to be noted later). If the belief is true, the action will succeed—or at least if it does not, the cause of failure will be something other than misapprehension of the circumstances.

Counterexamples can easily be constructed: false beliefs sometimes have serendipitous consequences. But relatively seldom—too seldom to negate the validity of the generalization, taken "for the most part."

Dogs can be deceived, they can size up situations wrongly, they can anticipate what does not come to pass, which is to say that they can have false beliefs. These mishaps, however, are usually of brief duration. Virtually all the beliefs of dogs are formed in action, so that their formation and their testing are contemporaneous; and if they fail the test, they cease to be beliefs. The principal if not only exception is the formation of a belief that some thing or situation is *threatening*, the effect of which is to motivate the dog to run away, retaining this belief indefinitely even though it may be groundless. For easily understandable evolutionary reasons, all animals have built-in biases to flee from potentially dangerous situations, and almost all novel situations are potentially dangerous. With this proviso, it can be said that the vast majority of the beliefs of dogs, and indeed of all subhuman animals that have beliefs, are true.

In this account it has been tacitly assumed that all the beliefs of beasts are categorical, that is, expressible (by us, who have language) in the form S *is* P or S *is not* P. This is of course not the case. The pelican diving for fish and perhaps catching one fish per twenty dives does not necessarily believe that there *are* fish in the bay, only that there *may be*. In logical jargon, her belief (or more exactly, the sentence expressing it) is *modal*. To take account of this complication, however, would be tedious, without requiring, as far as I can see, any substantive modification, at least at this stage, of the view of beliefs being presented.

SUMMARY

There are four ways of coping:

1. The plant way. Be built so that you can absorb what you need for growth and replacement of damaged or worn-out parts (light, water, carbon dioxide, various other chemical elements and compounds) when they come into contact with you. Have some means of distributing and collecting genetic complements. Acquire, if you can, by "slow learning," certain enhancers, such as water-conserving leaf structure; competition-discouragers; tropisms, especially for light; pollinator-attractors; forager-discouragers (e.g., thorns); seed-distributor-attractors (tasty fruits). But *doing* anything to go beyond these procedures to take advantage of transient opportunities is beyond your capability. You therefore have no need of internal indicators of external conditions; a fortiori you have no beliefs.

2. The way of the brainless animal: reflex. The sea-anemone is very like the Venus fly-trap, though slower. Jellyfish sting whatever comes within their tentacles and ingest it; their motility enables them to cover more territory, but not to hunt, nor to evade either; they have to rely on inedibility and sting. Sea stars are marginally cleverer: they can envelop mussels and pry them off rocks. They have one sense only, that of touch; no distance receptors. Some sessile animals, for example, barnacles and tapeworms, perhaps have no senses at all but are mere living filters.

3. The way of instinct. The animal has a repertoire of responses to situations, is programmed to execute the behavior most likely to further its reproduction or nourishment, as well as that of its offspring and in some cases its close relatives. This necessitates finer sensory discriminations and the sizing-up ability. Examples: migrating, mating, web-spinning, nest building, feeding of young (and the behavior of the young receiving the food); all building except by hominids, it would seem, even beaver lodges. Apparently all behavior of arthropods is instinctive. Even much human behavior is instinctive, that is, unlearned, executed by the whole organism, and non-teleological (from the point of view of the agent).

4. The way of choice. Acquisition of memory enables an animal to modify its own behavior patterns, by trial-and-error learning. With memory

comes the notion of a sequence in time, therefore of *consequences* of happenings and doings, therefore of a future, therefore of *goals*. Goals are objects of *desires*. Hence animals with memory can *choose* between courses of *action*, the anticipated consequences of which differ in *value*, imagined desirability.

Reflex and instinctive behavior can proceed when stimulated directly by sense-transmitted information. Choice, however, requires anticipations as to how alternative courses of action will modify the animal and its surroundings, and subjective comparison of the relative values of these projected modifications. Deliberate chosen behavior is teleological, motivated (not merely stimulated) by desire.

Anticipations are one kind of *belief*: true if things turn out that way, false otherwise. Thus beliefs are aids to coping in motile animals whose behavior is more than merely reflexive and instinctive. They occur at the level of action—non-instinctive, learnable behavior.

Beliefs arise only in connection with coping, at any rate in the beginning—that is, they are all practical to begin with, and are frequently jeopardized; success (surviving to reproduce) depends on having the right ones, those that survive these tests. This means that propensities to form erroneous beliefs get selected out. So nearly all beliefs of the kind subject to the selection process are true; that is to say, they constitute a sufficiently accurate map of whatever is the object of coping. The accuracy is practical, meaning that reliance on it does not (usually) lead to surprise, damage, or other embarrassment.

CHAPTER 3

❧

Language

Besides the mysterious faculties of laughter and tears, the most fundamental behavioral differences between people and beasts are three: language, fire making, and tool making (beyond hooks and poking sticks). Other activities universal among people but not shared with beasts, such as art, music, organized warfare, and superstition, are derivative from this triad, the members of which themselves were probably connected with one another in their origins.

Language, as Paul Grice has taught us, is primarily a means for getting other people to have beliefs that we want them to have, which, ordinarily, are beliefs like some that we have already. There are other such means, but language is incomparably the most efficient and versatile.

What sharply distinguishes language from warning cries, territorial markers, mating calls, and other communications between beasts is the employment of *arbitrary symbols* arranged in accordance with (also arbitrary) rules of *syntax* into *sentences*: structures which when heard (or read or otherwise perceived) by a person privy to the system convey a *meaning* to that person—the meaning, moreover, that the utterer intended to convey.

The dances of bees seem to satisfy these conditions, inasmuch as bees appear capable of communicating precise information to each other concerning one topic of crucial interest to them, the location of pollen-bearing plants, by body movements that have no natural connection to what they are about. These wiggles, then, may be true symbols, syntactically arranged. If, further, they are unlearned, then it must be the case that there can be such a thing as an instinctive language, even though of limited scope. It might be hypothesized that bee dances are an elaborate and sophisticated system of pointing. But pointing itself is a symbolic gesture, not found in non-human animals, unless turning of the head so that the nose "points" at an interesting feature is to count.

Leaving aside bees (and perhaps termites and other social insects), language as defined here is the unique possession of people. And though the ability to acquire language is innate, the development of the ability requires teaching.

The noises that beasts make are not arbitrary but instinctively produced in stereotypical situations. This has the advantage that every cat or blackbird can "understand" every other cat or blackbird; but the range of information communicated is bound to be very constricted: "Keep away from me," "Let's mate," "Predator in the vicinity," and a few more recurring and important messages exhaust the repertoire.

These "natural" signals, moreover, are not rudimentary *words*; they are *exclamations* which, being complete messages, are more like rudimentary sentences. That is why they cannot be combined, as words can, into other sentences to produce the virtually infinite varieties of meaning afforded by a true language.

Nevertheless, by making stereotypical noises, beasts manage to induce in other beasts certain beliefs about what is happening or about to happen: where progeny or prospective mates are to be found, what actions are likely to meet with resistance, and so on. That is to say, they manage to convey information. The fact that this can be done implies not only that the utterers of the noises are hardwired to produce them in (and only in) the circumstances in which they are appropriate (a certain noise—let us transcribe it as "Jaguar!"—is emitted when and only when a jaguar is perceived), but that the converse ability, to react to the noise "Jaguar!" in the manner appropriate to perception of a jaguar, is also hardwired. These are distinct abilities. Both involve meaning, but in different ways.

The second, the auditor's ability, is a particular exemplification of the sizing-up ability of higher animals, to interpret sense data, specifically, to infer unsensed features from correlated sensed ones: fire from smoke, prey animals from spoors. These are examples of what Grice calls "natural meaning." (Surprisingly, Wittgenstein's illustration of a simple "language game," the workman ordering his assistant to bring him this or that tool, might have had historical validity if, as has been suggested, tool making provided the impetus to language development.)

But does the animal *emitting* the sound mean anything by it? In many cases, certainly not; the kicked dog just yelps, the concealed hunter just sneezes; though other animals make inferences from them, their sources do not *intend* them to be thus picked up. So they fail Grice's test for "nonnatural" meaning, which requires the emitter to intend the sound to be informative, and the listener to recognize the emitter's intention. What, however, about warning cries? It is hard to resist the notion that the quail or monkey sounding off when it perceives a predator intends its action to cause its fellows to flee. But as we have seen, instinctive actions are not purposive, so if the warning cries are instinctive, as most if not all of them pretty clearly are, they too are devoid of meaning—just as the spider or the swallow does not *mean* to construct a trap or a depository for her eggs and shelter for her young.

Perhaps, however, on some rung of the evolutionary ladder sounds begin to be made intentionally. There is anecdotal evidence aplenty of (for instance) dogs barking and arousing the household when they smell smoke. After all, there is no reason why an act should not be both instinctive and deliberate. If that is the case, and if human language arose from a previously existing repertoire of "natural," instinctive cries—and it is hard to conceive of another possibility—then the transition from (say) instinctively uttering a vocable MOO! on perceiving an aurochs, to making the same sound deliberately *in order to communicate* the propinquity of an aurochs to one's fellow hunters, would be a natural one, and the noise in this case would satisfy the Gricean conditions for non-natural meaning. More hurdles, however, would remain to be got over. Onomatopoetic vocables such as MOO! have natural reference, so to speak; but their supply is meager. Lucy, or whoever, had to invent *arbitrary* symbols and establish their references culturally.

Finally, *syntax* had to be invented. MOO! and AXE! had to be transformed from exclamations—proto-sentences—into *words* from which further complex structures of meaning could be constructed ad lib with the aid of all manner of vocables, some of which—verbs and adjectives—refer to entities such as qualities, relations, and actions, which are not obvious discrete features of the sensible manifold; and hardest of all, the "joints" of sentences: vocables not referring to anything, nevertheless necessary for constructing fabrics of sentences: prepositions, conjunctions, and other linguistic "particles." The grammatical problem of advancing from atomic exclamations to molecular sentences is like the evolutionary problem of getting from single-celled organisms to multicellular plants and animals.

Pre-linguistically, let us suppose, MOO! signified aurochs, GURGLE! signified river. With these resources one could direct attention to an aurochs or a river; but to describe the particular state of affairs "There is an aurochs between us and the river" was beyond capability, unless, perhaps, supplemented by gestures. Lucy invented a cry NUS and somehow got her tribespeople to comprehend that when added to another cry it indicated that the object denoted by the latter was comparatively close; and still another noise FUS, "comparatively far." Then MOONUS GURGLEFUS would be the *sentence* she needed. Bypassing recourse to visual or other sensory input, the hearer of these noises would reconstruct in imagination the sizing-up that had been made by the utterer on the basis of perception. And this sizing-up would be a *belief*, an appreciation on the basis of which the hearer would be willing to act, at any rate in the normal case: for language will not work unless normally hearers believe what they are told. That is to say, the default state of the human psyche when a description of a state of affairs is input, is BELIEVE. When we are told something, we believe it, in the absence of some particular reason not to; it is not the case that we disbelieve or suspend judgment awaiting some further positive reason to believe. True, the acquisition of sophistication is largely a matter of learning to go counter to this natural tendency; but that fact only highlights what the natural tendency is.

Credulity has its dangers; later we shall dwell on them. Nevertheless, the tendency to believe what one is told should be ranked with the opposable thumb among the most precious peculiarities of the human race. It entirely revolutionizes education—indeed, makes it possible. Without it, an animal can acquire beliefs only the hard way, directly from its own experience; with it, everybody's experience can go into a pool from which everybody else can draw. While it is true, no doubt, that all knowledge begins in experience, in fact the vast majority of a given human being's beliefs come not from his or her own perceptions but from being told, and the more educated the individual, the greater the percentage.

Each puppy must form its personal conception of a cat from its own encounter with a beast apt to hiss and scratch. A child, on the other hand, may be told by his parents that there are long green leathery scaly toothy beasts called crocodiles lurking by the river, and that he is to flee when he sees one. Thereby the child forms his personal crocodile concept, and action-guiding belief, in advance of any actual encounter. It may happen that a day, a month, even a lifetime passes during which no crocodile is met. The belief nevertheless persists. It is not checked by the believer, it is not rubbed up against reality to find out whether there really are crocodiles, or whether they are as dangerous as they are purported to be. This is something that never happens to the languageless animal, in which an unchecked belief can hardly arise, let alone persist past the specious present.

Being told, moreover, is the *only* way in which certain kinds of beliefs can be acquired, such as concerning events that happened elsewhere or before one's birth. And thus language bursts the bonds of the specious present. *A* can get *B* to form expectations extending indefinitely into the future: "Godot will come *some* day." The cycles of day and night and of the seasons acquire names, and events can be associated with them *as* past and *as* to come.

The Nevada work with chimps suggests that they already are mentally equipped to have a true language: they can learn it and, we are told, reteach it to their young! So perhaps the leap from cries to language was not over so wide an abyss as one tends to think it must have been.

Finally, and portentously, language makes *lies* and *stories* possible.

Pre-linguistic sizing up can be done wrongly, whereupon the animal's appreciation of the situation in which it finds itself is mistaken. But the error is confined to the individual and seldom persists; the defective sizing up is soon followed by action that reveals its deceptive character. If, however, the mistaken appreciation has been transmitted by speech to another person in the interval between error and test, the recipient may continue to believe wrongly after the matter has been cleared up for the initiator. And of course the technique of lying, the deliberate inculcation of misinformation known to be such, must have originated early on: Cain, the first murderer, was also the first liar.

But the very possibility of lying presupposes an ability to imagine a state of affairs while *dis*believing that the state is actual, or—a further complication—neither believing nor disbelieving its actuality but merely entertaining the possibility. In this way imagination assumes an activity of its own: it becomes possible for a person to think of "what it would be like if…," turning imagination into the "creative" faculty. One who imagines a state of affairs or sequence of events in this way may narrate the product, thereby telling a *story*.

A story is not a lie, nor is it misinformation, if its nature is understood by its audience, that is, if the listeners grasp that they are not (as in the "normal" case) to believe what they are being told, but merely to imagine, along with the storyteller, "what it would be like if." But thus to hear a story *as* a story is a sophisticated—one might say unnatural—performance and very liable to go wrong, as we know from countless experiences in our own time. Stories, pure products of imagination, *escape* and become beliefs not distinguished, as beliefs, from sober and reliable testimony. They become *legends*, objectively speaking, but in the subjectivities of the believers, *histories*. Since they are about the past, the possibility of rubbing them up against reality seldom exists, at any rate in preliterate societies.

The upshot is that language extends the scope of imagination enormously; this in turn makes possible the formation of beliefs that may persist even when opportunities to check them against reality are indefinitely postponed.

Moreover, a story often contains, or is, an *explanation*—"How the Leopard Got his Spots" and the like—and as such may be believed to represent fact, even by its inventor, much as a scientist of today, not too puritanical in his or her methodology, may believe his theory as soon as he has invented it and before he has put it to any test. Indeed, legends *are* primitive theories.

There is a basic human propensity to find order in things, and to put it there if it is not apparent of itself: witness the naming of constellations, schemes of divination, mnemonic verses. Explaining things is one way of putting order into them. And in primitive conditions, the known behavior patterns of fellow human beings afford the only available models for ordering the world. What is lightning? It is an expression of anger—the anger of a powerful being up there. Forthwith the being is imagined and named Zeus.

Was whoever first imagined Zeus telling a story in our sense, a conscious fiction with suspension of belief? Not necessarily. Zeus explained lightning. There being current no alternative explanation, the inventor of Zeus thereby ordered lightning into the scheme of things. There was nothing to countervail the natural tendency of people hearing the story to believe it, to take Zeus as a real denizen of the one world of their experience. What is more, there is no reason why the inventor himself or herself should not have believed it. Thus the legend of Zeus could have been born without conscious deception, or even confusion, on anybody's part.

But the communication of information and misinformation, in utterances that can be said to be true or false, that is, representing or misrepresenting how things are, is by no means the only thing that can be done with language. There are, besides, the multifarious *speech acts* to which philosophers beginning with J. L. Austin and John Searle have devoted so much attention: promising, commanding, marrying, judging, voting, bequeathing, and so on. The utterances constituting the acts are neither true nor false, but, like other actions, to be adjudged as felicitous or infelicitous, appropriate or inappropriate, prudent or rash, and the like. For example, *commands*. Pre-linguistic bands are hierarchically organized into pecking orders, in which those higher up can prevent the humbler from doing certain things, but there is seldom a way to convey the intention involved in a positive command or request, save in a few stereotyped situations. With language, an internal representation of the state of affairs to be brought about can be produced in the person addressed, who straightway obeys (the command was successful) or else doesn't (unsuccessful). This power bestows a tremendous evolutionary advantage on bands possessing it, which can for the first time be efficiently organized to carry out common purposes.

Non-representational speech acts being neither true nor false, they are not directly objects of belief. Nevertheless, they often, perhaps always, presuppose beliefs; and inasmuch as they constitute the fabric of society (as distinguished from mere flocks, herds, gaggles, and bands), we must attend to them. But later (chapter 20).

SUMMARY

Communication between animals is primarily a means for getting one to acquire a belief that another wants him or her to have. A natural vocable, such as a warning cry, causes the recipient to imagine the situation encountered by the utterer and to react suitably. But a true language requires more than that: the "non-natural," arbitrary symbols, syntax, and semantics from which new *sentences* (as opposed to stereotypical exclamations) can be formed ad lib. This achievement opens up a new way of forming beliefs: by being told, enormously increasing the scope of beliefs for people.

Since for this to work the default state of the receiver must be BELIEVE, it opens up the dangers of lies and misinformation also. And *stories* can be told: harmless if judgment is suspended, but apt to "escape" and become legends and received explanations. Thus unlike beasts, people can believe for indefinite lengths of time what has never been rubbed up against reality, and is in fact not true.

Besides information, language makes possible the *speech acts*—commanding, promising, marrying, and the like—of J. L. Austin, which in turn create the *institutional facts* that largely constitute the structures of human societies.

CHAPTER 4

✧

High and Low Beliefs

Some beasts have memory, for they can modify their behavior in the light of experience; and they have imagination, for they can anticipate what is to come, as is shown by their ability to act intentionally. Beastly imagination, however, is very shallow, limited to the short term, as we have seen.

In compensation for their constricted range, nearly all beliefs of beasts are true. All of them are based either on the animals' own direct experience or on the limited repertory of signals current in their species. The former result from wired-in sensory capabilities honed by evolution to provide accurate information which the animal efficiently sizes up; the latter are seldom deceptive. Furthermore, a beast's belief is not formed idly but for use in directing immediate action. Consequently, if it is false, the animal is likely to find this out at once; things not being as it supposed them to be, the behavior predicated on the belief will in most cases not attain the goal sought; the belief, we shall say, has been rubbed up against reality and found wanting. The false belief will be extinguished, and perhaps the believer also. Thus while particular beliefs are not heritable, any tendency to form beliefs that depart very far from the way things actually are will, in general, have negative survival value; other things being equal, the possessor of this trait will be less able to cope. Of course some mistaken beliefs do not matter in the end, and some even have serendipitous consequences, such as Columbus's underestimate of the earth's circumference. But the general rule is obviously valid.

Because people have language, which enables beliefs to be formed other-wise than through direct experience of their objects, the range of human belief is vastly greater than that of beasts. The beliefs of all persons speaking the same language form in principle a pool to which each member of the linguistic community has access; and given that every language can be translated into every other, this community embraces—in principle and increasingly in practice—the entire human race. The beliefs of an individual member even of a small and isolated hunter-gatherer band that are based on that individual's personal experience are but a fraction of all the beliefs he has; this fraction becomes almost vanishingly small in the case of an educated person in a literate culture. That is to say, most of what we believe is what we have been told or have read about. Moreover, the connection of our beliefs with our coping is in the vast majority of cases merely potential, in contrast to the immediate practicality of beastly belief. We all believe that the climate of Patagonia is cold, and especially so at the times of year when it is hottest in the northern hemisphere, though the need seldom arises for us to take this fact into account in planning what to pack.

We recognize too that—again in contrast to beasts—people (except me and thee) are prone to believe many things that are false. "To err is human; not to, animal," as Robert Frost noted. A principal purpose of this book is to explain why this is so. Two preliminary points: first, that this fact presents a paradox; second, that pessimism about it can be overdone. Both are basic to the explanation that will emerge.

The paradox: As we have stressed, having beliefs is a necessity for the coping of all animals except the simplest, and to be of service they must be *true*, that is, they must represent things as they really are at least in those respects pertinent to dealing with them. We have taken it as obvious that evolutionary selection tends to maximize the reliability of beasts' beliefs, to the point where hardly any of them are false. Yet the case of human animals, the most successful of all mammalian species according to the survival-and-reproduction criterion, is exceptional, and a big exception. Indeed the saying is true—as everyone believes on the basis of personal experience—that *Mundus vult decipi*, people not only do but *want to* believe what is false.

A pessimistic belief, which, however, should be tempered by realization that probably most human beliefs *are* true, do really represent things as they are in pertinent respects. We are at no disadvantage relative to the beasts with respect to the truth of the beliefs that we form on the basis of our individual experiences; and most of what we are told—for instance, about the climate of Patagonia—is true too. The information in the *Encyclopaedia Britannica* is overwhelmingly reliable. You (probably) and I have not been to Patagonia, but somebody has, and has accurately reported what it's like there. In our terminology, "Patagonia is cold" is a belief that has been rubbed up against reality, has passed the experiential test: visitors found ice on puddles, had to bundle up when they went outdoors, and so on. Why do we believe that this is so? Perhaps only because of our instinctive tendency to believe what we are told. But that, as we have noted, is a good reason, in the sense that one who acts on it will have more successes than failures. And we are better off than that: we can point out that we learn from experience that travelers' tales, while often inaccurate, seldom depart from the truth on gross points such as the nature of a climate unless the tale-tellers have some personal interest in propagating falsehoods; and that in a case like this, it is hard to conceive what motive could exist for misrepresentation, not on the part of one explorer but unanimously among dozens, at least, who must have been there and reported on it.

Testimony is *generally* reliable because people *generally* size up with *fair* accuracy the data which their sense organs derive from their surroundings—this is an accomplishment we share with the beasts—and because *more often than not* witnesses neither deliberately nor inadvertently falsify their reports. It is against this background of veridicality that the problem of false believing is to be discussed.

Having made these two points—that, paradoxically, false beliefs are common among and only among the most successful mammals, and that, nevertheless, they do not predominate—let us begin the inquiry after the reason for this state of affairs.

The first thing to note is that while beasts' beliefs have only one source, the individual experiences of the believers, people's have three: individual experience, testimony, and imagination. As we have just seen, testimony on the basis of experience is an extension of experience, but does not introduce an element of a radically different kind. It is continuous with, and probably arose out of, the kind of information gregarious beasts convey to one another through signals. To be sure, testimony even when offered in good faith is less reliable than direct experience, for linguistic expressions seldom convey the contents of experience more than approximately and are subject to further errors in interpretation at the receiving end; and testimony is not always offered in good faith. These considerations on their face account in part for the fact that "to err is human." That is, we should expect a priori that animals with language should be more prone to false belief than the languageless. This would be in itself an evolutionary disadvantage, which, however, would be more than compensated for by the vastly enlarged capability of forming true beliefs given by language. Moreover, false beliefs are hazardous to survival only insofar as they concern matters of practical import, where misapprehension endangers the efficiency of coping behavior. Outside these areas, false beliefs would be harmless. If, when a tiger is rustling the grass ahead of you, you believe that an aurochs is making the noise because you have been told so by your brother-in-law who secretly hates you, you are in trouble; but if you believe junk mail from a real estate speculator telling you that Patagonia is a tropical paradise, the belief has no practical effect as long as you have no money to invest or with which to buy a plane ticket. Beasts are exempt from liability to this kind of error because (with few if any exceptions) they cannot form, or at any rate retain, beliefs that have no practical consequences; their powers of imagination are only rudimentary.

Such is far from the case with people. People have stocks of words, and a command of syntax by virtue of which they can be joined ad lib into internal representations that need correspond to nothing in experience. In contrast to cats and dogs, which can perhaps imagine states of affairs literally one jump ahead, people can imagine anything that their language is capable of depicting; and if there are any limits to that, we do not know what they are.

You and I easily and sharply distinguish between what we believe and what we only imagine. We imagine winged horses, mermaids, time travel, and so on—that is to say, we form internal representations thus describable, but we by no means believe there are, or even could be, any such things in reality; we do not assent to these representations. And we can amuse our sophisticated friends by telling them stories about such pseudo-entities. If our young children overhear them, however, their instinctive tendency will be to believe them, as we have noted: to assent to them, to judge them things that should be taken into account when planning their actions in the real world. In this and other ways such as deliberate bamboozlement, mere constructions of imagination get transformed into beliefs. This is a source of belief distinctive of our species; no other animal possesses the capability of forming a belief independently of *any* experience of what it is about.

HIGH AND LOW BELIEFS

The beliefs of beasts consist of memories and expectations. A beast coming to an entirely novel situation may be able to cope with it, but only by way of instinct, reflex, or luck, none of which is belief. A kitten must confront an actual mouse—must see its gray form, hear its squeak, smell its mousy odor, observe its rapid scuttling, taste its blood, in Kant's ponderous but apt terminology "synthesize the sensuous manifold"—give a unity to this congeries of sense data and store it in memory so that when, subsequently, a similar experience occurs, the animal sizes up the situation as containing a mouse, that is, an edible object worth pouncing on. This (or something of the sort) is the kitten's mouse concept.

Although we do not know what it is like to be a cat, we do know that when a cat encounters what we call a mouse, there is some occurrence in the cat's psyche that stands in some representative relation to the mouse. It is in light of this fact that I venture to speak of "the cat's mouse concept." This expression, however, may be misleading. It is naive to suppose that cats must divide the world into "objects" which would be denoted by nouns if only the beasts had a language. Possibly cats' concepts, being strictly practical in origin and use, are primarily gerundive, that is, would be more perspicuously rendered by "to-be-pounced-on" (including what we call mice, moths, and balls of string), "to-be-scratched" (trees and sofas), "to-be-hissed-at" (dogs, other cats, and so on.) *We* think: There's a mouse, we must set a trap. The cat perhaps skips the "demonstrative" step.

On the basis of this concept, next time around, the kitten *believes* that what it sees is a mouse, that is, expects that it can be pounced upon for fun and profit. The kitten may be mistaken, landing instead on a baby armadillo or a wind-up toy. If it does, it will be frustrated and surprised, but constructively: it will refine its mouse concept so that its subsequent beliefs as to what are trustworthy signs of mice will be more accurate.

The kitten, forming its mouse concept and its beliefs about mice, has done so in an encounter with an item of reality, namely, the mouse. Let us call any belief formed this way, in an encounter with the thing or situation that the belief is about, a *low belief*; and any belief formed in some other manner, a *high belief*.

As we have seen, all the beliefs of beasts are low, because the only other ways of acquiring beliefs are by being told or by the exercise of imagination, which only people can do. Being told and imagining are, of course, not encounters with what the resulting beliefs are about.

But what are beliefs about? Every belief is about something, and if the belief is low, something that exists.

For three reasons, low beliefs are not necessarily true (that is, representing their objects satisfactorily enough for successful coping), even though by definition they are acquired only in encounters with items of reality that they are about. The first is that all creatures are subject to error; they may misperceive and misremember things. A dog may think (yes, dogs think) that a wadded cloth at a distance is a cat; your or my encounter with a wisp of fog in a graveyard may be thought to be with a ghost. Further experience, however, reduces the error rates of most animals. Second, Nature can always, on her part, spring surprises; expectations of her future behavior may be unfulfilled no matter how well grounded they are. There is little that animals in general can do to cope with earthquakes and tsunamis.

The third reason why some low beliefs are untrue is of more theoretical importance: it is that a belief may be both false and *right*.

To speak in (I hope) harmless metaphors, Nature does not care about what her creatures believe, only that they *do* the *right* thing, to wit, whatever best promotes coping—surviving and reproducing. Inasmuch as what they do depends on what they believe, she has seen to it that they tend to have the right beliefs, to wit, beliefs that lead to right actions. Ordinarily, right beliefs about the things with which an animal must cope are true beliefs; that is why the creatures have been provided, through evolution, with exquisitely sensitive and accurate sense organs, and the software to organize the inputs from them into accurate representations. But "right" and "true" are not synonyms. It is a contingent fact that having true beliefs is highly correlated with doing the right things. A "better safe than sorry" policy of believing that any novel situation is threatening and hence fleeing from it, may be evolutionarily sound, even though many such situations are not in fact dangerous. Thus beliefs involving fear or fright, which are expectations of pain, stimulate avoidance behavior, which is likely to preclude further investigation and correction. In such a case, an initial perception or sizing-up which may not have accurately represented the facts will persist even for a lifetime, a false (though low) belief biasing the animal's attitude toward heights, open spaces, social workers, or what have you—things in themselves harmless or even beneficial. When beasts hold such beliefs, however, they are bound to be idiosyncratic, because they arise out of one-time incidents in the experiences of particular animals, and are relatively rare. By definition, evolution favors beasts with genetic characteristics that evince themselves in ability and tendency to form right beliefs; and in fact these characteristics are very largely the same ones that produce true beliefs. Consequently, it is safe to say that *almost* all the settled beliefs of beasts, especially those of mature individuals, are true.

BELIEFS OF PEOPLE

No one would be tempted to extend the preceding generalization to include our species. Looking at the human propensity for bizarre beliefs from the evolutionary standpoint, it may seem not only lamentable but paradoxical. If the coping strategies of animals are based on their beliefs, is it not evident that, other things being equal, the more accurately an animal's beliefs represent the way things really are, the more successfully the animal will cope? Yet the human being, the most successful animal by far, is massively and uniquely susceptible to holding false beliefs. How can this be?

Briefly, the paradox is resolved when we acknowledge that people, because they have language, acquire most of their beliefs differently from the way in which the beasts do; and that many of the beliefs so acquired are either not bound up with coping, or if they are, are bound only in indirect, even devious ways.

People are animals, after all, and can, and to some extent do, acquire beliefs the same way beasts do: from experience of novel situations. This is of course true of many crucial experiences that provide us with our basic background of how the world goes (what is solid and what not, eye-hand coordination, no rain from clear sky, etc.) and take place before language acquisition at about age two. After that time, experience becomes more and more socialized: we are told beforehand what to expect and this information colors the experience, when it comes, in various ways.

Among beasts, not only are all beliefs low, but all are *individually* low: each one is acquired by the individual concerned from its own experience. Beastly beliefs are not, so to speak, contagious, except in the very limited sphere of information transmittal by signals, where, significantly, fear is almost always involved. When we consider children, however, we need to distinguish between *personal* and *social* low and high beliefs. A personal low belief is one that has been acquired by personal experience—for example, that fire burns by being oneself burned. In any human group there are likely to be some who have this belief, acquired this way. They warn their fellows who have not yet personally been burned but who heed them and thereby acquire personal *high* beliefs that fire burns. For those of them who subsequently get burned, their personal high beliefs become personal low beliefs. But if anybody in a society has been burned and expresses to others the belief that he has acquired thereby, and the veridicality of his or her account is generally accepted so that everybody believes it, the belief is a *social low* belief from then on. Henceforth in this book, the unqualified expressions "low/high belief" will refer to *socially* low and high beliefs. The context should make clear what society is being talked about; usually, at least for low beliefs, it will be the whole of humanity, as in the example just given.

Although each beast must acquire all its beliefs for itself, each winds up with virtually the same ones as all its other fellows have. People's low beliefs exhibit the same unanimity. Everywhere and at all times people believe that rain falls only from cloudy skies, two and three are five, water quenches thirst and fires, top-heavy structures are unstable, insults create enemies, most men are bigger and stronger than most women, fingernails grow back when cut but fingers don't, fermented drinks make you drunk. Myriads of such boring beliefs, seldom traceable to their sources in experience (either individual or social), seldom even consciously formulated, constitute the indispensable core of directions for getting around in the world. They are genuine beliefs: intelligent immigrants from Titan should not be expected to have very many of them. Some low beliefs, to be sure, are not universal: not all tribes have to cope with crocodiles; but all those that do have the same expectations and adopt similar strategies and precautions.

Is it right, though, to call such items beliefs? Are they not, rather, basic *knowledge*?—Yes, they are knowledge, but not "rather." They are beliefs, and they are true, and true belief = knowledge. (This equation has been rejected by virtually all philosophers since Plato. See, however, Crispin Sartwell's unjustly neglected articles.)

Low beliefs of both beasts and people being nearly all true, it follows that nearly all falsehood pertains to social high beliefs. Yet a society's high beliefs are its most prized possessions, to be protected at all costs. In what follows we shall see how and why this is so.

SUMMARY

Virtually all the beliefs of animals without language are formed in encounters with what they are about. Let us call them *low* beliefs. And with few exceptions they are *true*, that is, they represent the way things really are with sufficient accuracy to further coping with them, because they immediately get rubbed up against reality, whereupon if they are false, the animal abandons them. People too have low beliefs, which like those of beasts are overwhelmingly true and the same for all human groups. But people can acquire beliefs in other ways besides encounters with what they are about: by being told (testimony) and by imagination. Let us call these beliefs *high*.

If people in some community have acquired low beliefs and transmitted them to others who have not had the encounters generating them, these beliefs are individually *high* for those who take them on testimony, but *socially low*. Henceforward in this book the adverb will usually be omitted; "low belief" will mean "socially low belief" unless the context indicates otherwise.

It is paradoxical that the most successful mammalian species, *Homo sapiens*, should be the one harboring the vast majority of false beliefs. Much of the rest of this book will be devoted to explaining how this can be and what are its consequences.

CHAPTER 5

༄

The "Will to Believe"

Another difference that language makes is that while beasts have no control over their beliefs, people do, to some extent.

To say that beasts cannot control their beliefs is not to say that they are doomed to believe everything they sense. The savvy coyote, casing the henhouse and on the lookout for traps, is not easily taken in. But that is because by experience he can discern features of the situation that are "not right" and restrain him from straightway concluding that the tethered hen is free lunch. His sizing up is done skillfully: pertinent evidence is not overlooked; the significance of details is properly evaluated. He exercises judgment. But the integration of all the factors is assented to, becomes belief, automatically: it tells him that there is a trap. If starving, he may risk it anyway, in spite of his belief; he does not first "talk himself out of" (significant phrase!) his belief. A coyote has no "will to believe."

He could not have. Nature has endowed him with his acute senses and his judgment so that his internal representation, on which he bases his life and death decisions, will indicate how things really are. This purpose would be frustrated if the question of accepting that representation were left to his will—if he could accept it or reject it according to his wanting things to be one way rather than another.

The same is true, or nearly true, of people where the type of belief is that which we share with the beasts, namely a personal low belief. Nobody nohow (*pace* Orwell) can be got to believe that 2 + 2 = 5. Nevertheless, the prevalence of wishful thinking (in people other than me and thee), which is nothing but believing what we want to believe, exhibits yet another ability that sets us aside from our beastly cousins.

With certain exceptions which we will take up later, the beliefs that we *decide* to accept or reject are beliefs based on testimony, that is, they are personal high beliefs. At this stage of the history of science, the rotundity of the earth, the basic principles of Darwinism, and Gresham's Law qualify as social low beliefs. But direct experiential evidence for them is vouchsafed to few; most take them on authority, and some, relying on other authorities, still reject them. William Jennings Bryan famously denied that man is a mammal. He did so because he had *faith* in the Bible, which, he (erroneously) thought, asserted the contrary.

THE PHENOMENON OF FAITH

You can believe testimony for the reason why it is inherently plausible; or because you believe the witness is trustworthy; or for some combination of these reasons, including combinations with negative scores on either, though not on both.

There may be good reasons for believing a witness. If what the witness tells you is confirmed by you—that is, you have an experience whereby it becomes your personal low belief—that is good evidence for trustworthiness of the witness; and more such confirmations increase it. Your belief that the witness is trustworthy becomes your own low belief, but with this difference from low beliefs in the workings of nature: with nature, induction works, if you take proper precautions; with people, it can't be counted on in the same way. If Jones has always told the truth, the inference to "He will tell the truth next time" is not a straightforward induction like the sun rising tomorrow; it is via the intermediate step, "He has shown himself trustworthy up to now," which does not validate the next step the same way. One must take account of his motives, the inherent plausibility of his story, possible bias, and other such factors.

But faith in matters of belief is usually a corollary of faith in matters of other personal relations. Troth, truth, trust are all the same word basically. The giving of untrustworthy testimony is a hostile act and therefore not to be expected of or imputed to those with whom one has positive human bonds.

The default state with respect to testimony is BELIEVE. This is particularly the case with one's immediate family (parents a lot more than siblings, though). One is innately biased to accepting their testimony unless there are cataclysmal reasons for rejecting it; and such rejection is traumatic.

This brings in the transitivity of faith: one has faith in parents (or parent surrogates: whoever one has bonded with) and therefore with those in whom the parents have faith. And since in the forty-member band there will be pressures in the direction of conformity with a standard story for the band, one will have faith in the shaman, or the tradition, or whatever the high beliefs of the band connect with.

This is, so to speak, natural faith, and it has nothing to do with the will. One just finds oneself having these beliefs. They have "positive affect."

As one forms relationships with other people, one may accept their beliefs as well; or their having the right beliefs may be a basis of a relationship. The main point is that faith in a person comes first, acceptance of that person's beliefs, second. But this is a very important difference from the beasts; there is no possibility, with them, of any such linkage of believing with bonding, even though beastly bonds may be just as tight as people's.

This still entails no "will to believe," but it may be getting close. It begins to show that there may be motives for believing, and where motives operate, will does too.

Marxists and deconstructionists hold that all important beliefs are determined by motives (except, of course, for their own). All believing is wishful thinking. This is too extreme. The question is whether *any* believing is literally subject to the will. And if so, how?

One should eliminate from this discussion William James's example of the chasm jumper. That is not a matter of whether one can believe that *p* at will, but of self-confidence. Of course one can "believe in oneself", in a broad sense, but it is hardly propositional. 'I can make it across that chasm' is too far removed from 'I am a descendant of the Marquess of Montrose.'" From "It would be nice if I were a descendant of the Marquess of Montrose" one can't go directly (unless one is mad) to "I am a descendant of the Marquess of Montrose." But if there is a family tradition, or portraits show an uncanny resemblance, one can. And—this is the more important point—one can not only assign considerable weight to flimsy evidence but one can also deliberately downplay contrary evidence, ignore gaps in it, and so on. Consider A. D. White's relation of the process of canonization of St. Francis Xavier. Or the genesis of the Chanson de Roland.

It is a common experience to find when reading a newspaper account of something of which one has personal knowledge, that they have got it wrong—factually—and to reflect that this probably goes on in many cases, where, however, one just goes along with the reporter, not raising any questions.

Since true low beliefs testified to will check out, and high beliefs testified to aren't controverted (by definition), it's all too easy for a witness to acquire a virtually perfect "inductive" record. False low beliefs are the only hazardous area for the liar.

It is time to define faith. To have faith in P = to accept the testimony of P without further corroboration and notwithstanding contrary evidence.

One acquires faith in another in much the same way as one falls in love; indeed, it may sometimes be the same process. How is will involved? Is there a Will to Love? Etymologically it would be very close to a Will to Believe.

It is a plain fact that one can talk oneself into (or out of) this or that belief. Only, however, against the background of a *Weltanschauung*, a belief field, and only where we have a "live option"—some antecedent attractiveness in the belief.

CHAPTER 6

✢

Eden

The apes that invented language had to be gregarious: chimpanzee-like rather than orangutan-like. They gathered edible plants and hunted small animals for food, which they shared to some extent. The bands were hierarchical, with a recognized (if sometimes challenged) leader. About forty individuals, plus their young, made up a band: more would have been unwieldy, fewer too vulnerable. They made things, including tools for making things, and they made fire. These three accomplishments—language, tools, fire—definitive of the genus *Homo*, were in place a million or more years ago, fifty thousand generations at least.

The next great leap forward, agriculture and domestication of animals, did not take place until ten or twelve thousand years ago—five hundred generations more or less—and its consequences spread slowly. Thus about 99 percent of human evolution has gone on in the context of the hunter-gatherer band, which is entitled to be called the "natural" human milieu, in the sense that we are still primarily adapted to it. We shall refer to this way of life as "Eden."

Let us consider the situation regarding high and low beliefs in Eden.

The low beliefs will include all the know-how needed to cope: these plants are nourishing, those are poisonous; those things are aurochs droppings; this is the way to make arrow poison; those reeds are the best basket material. Worldly wisdom: flattery will get you somewhere; your enemy's enemy is your friend. Mathematics and logic: two rabbits and three more rabbits make five rabbits in all—or even, tremendous feat of abstraction, two anythings and three more anythings make five anythings; he is either lying or he is telling the truth; since it rained last night, there will be water in the water hole this morning. Particular facts about the band's territory; the succession of the seasons; the habits of animals; orientation by stars. And so on and on: the band's *lore*. Items of lore—even mathematical lore—are miscellaneous and unconnected. If connections are suspected they are apt to be merely imaginative or coincidental, for example, between the lunar and menstrual cycles.

Although the way to prepare arrow poison may be a guarded secret revealed only in an impressive ceremony, most low beliefs are humdrum. People do not get excited about such matters as that wet clothes hung out in the sunlight will dry. Indeed, most such beliefs will not even register in consciousness.

Searle calls items such as these "the background"; Heidegger lumps them together as "being-in-the-world." Though emphasizing their importance, these philosophers seem reluctant to regard them as *beliefs* at all, thinking of them rather as some type of lower level mental entity. But hanging out the laundry is an *action* just as much as going on Crusade, and just as much a consequence of beliefs.

Most low beliefs will be true—again, in the minimal sense of enabling the believer to cope. They are not true by definition, for it is possible to continue to hold exploded beliefs. But where life is hard, the perilous consequences of the policy will manifest themselves in a fairly short run.

Let us turn now to high beliefs of long standing in Eden. Some will be high because intrinsically they cannot be rubbed up against reality: beliefs about first and last things, how the world began, what awaits the soul after death. Others are high because technology for checking them does not exist (the earth is flat) or they are about the distant past (the Chief is descended from the Sun) or it would be too difficult to test them (gods dwell atop Mount Olympus) or testing them is too dangerous to undertake (love-apples are poisonous; the Chief's shadow will kill anyone who steps into it). The first kind ("Some...") of untested beliefs mentioned here correspond to the "pseudo-propositions" of the logical positivists. But the second kind ("Others...") illustrate how a belief can persist untested for centuries while still being quite testable, and meaningful by any criterion. There is nothing *formal* about the notion of the high belief, nor about the proposition that expresses it. Yet these *are* *beliefs*—internal representations, assented to, of the way things are (or were or will be): pictures that the believers expect to be realized in their experience should the appropriate circumstances come to pass: in case one dies, or is transported (never mind how) to the summit of Olympus. Once the conditional-contrary-to-fact frame of mind has been created, the freedom from time that language affords even makes possible genuine beliefs about the remote past. One cannot have expectations about the past, to be sure, but one can form a conception of what one would have experienced had one (*per impossibile*, nevertheless imaginably) been present at the founding of the Chief's line or at the creation of the world; and one can assent to it.

Unlike its low beliefs, the high beliefs of each band will be different from those held in other communities not in close cultural contact. This will be so because high beliefs, like the languages in which they are expressed, are compounded from arbitrarily chosen elements.

Nevertheless, virtually all high beliefs, at least in primitive societies, will have two features in common.

First, since most high beliefs originate in stories, and all stories are about people and their doings—Raven, Moon, Little Engine That Could, Avarice, and so on, must be personified to become protagonists—all agencies involved in high beliefs will be persons or personifications, and their modi operandi will be human.

Second, high beliefs will tend to be *edifying*—supportive of social values.

Every animal in sizing up its situation out of fragmentary and jumbled sensory inputs must try to make sense of its experiences, and man most of all. Making sense consists in assimilating the unfamiliar to the familiar. Now, most low beliefs, though true, are intrinsically unconnected; they express how things happen, but they are dumb as to why they happen that way. Human action is the only element in primitive experience that exhibits any connectedness: the only *system* of interrelated doings that people have any grasp of is what goes on in their society. Hence, making sense of the world at large must consist in conceiving it in social terms: purposes, motives, will, good, evil, love, hate, and other emotions. Primitive world-views *must* be teleological; agency *must* be personal, volition *must* be the only causation recognized.

What goes on in nature then will be understood through stories of the deliberate choices and actions of beings that are persons though not human: personified animals, rivers, and the like, or invisible personal "spirits" (i.e., breaths, life forces): *demons, gods*. To them *guilt* can be ascribed for whatever is extraordinary, mysterious, and important for well-being—weather, famine, earthquakes, tempests, infertility, and all the host of things and events on which human existence is dependent but which are beyond the primitive human being's control.

The idea naturally develops that these powerful personages may be amenable to some of the influences that serve to modify behavior of one's fellow human beings: threats, entreaties, flattery, promises of reward. Such stratagems will be tried, and sometimes they will "work": the virgins are thrown into the volcano, and behold, the eruption stops; the happy outcome is taken as confirming the efficacy of the procedure. If it had not worked—well, attempts to placate the elders of the tribe often fail too. *Religion* gets established. (*Magic* too: its distinction from religion is superficial, being merely that the priest entreats, the magician commands.) A society's stock of high beliefs, and the beliefs involved in what we call its religion, are nearly coextensive. But not exactly: there are political and professional high beliefs, and a miscellany including once universally held beliefs such as the flatness and fixity of the earth.

There is no limit to the variety of forms and characters imputed to gods, nor to their relations with human beings, with one exception: they are never thought of as *indifferent* to human concerns. (The fact that Epicurus, Spinoza, and Deists such as Voltaire conceived of god(s) as indifferent to us is precisely why they were deemed atheists, and rightly.) Primitive common sense more often pictures them as menacing than as benevolent, indeed as operating a sort of cosmic protection racket. And like Mafia godfathers, they are never unapproachable: pay their price and they will at least leave you alone, at best aid you in your endeavors. But *only* aid: they will never do anything for you that you could do for yourself. It is easy to see why these restrictions obtain. Belief in indifferent gods would be unedifying. For paradoxically, the prime benefit that the believer in even the most horrible deity derives from his or her faith is *hope*. Primitive people supplicate the volcano god as moderns pay fees to quacks, because doing *something*, no matter what, seems preferable to giving up, acknowledging that there is no help to be had. And a tendency to believe that gods might be so amiable as to spare you from the effort of exertion would render the believer inactive and, if persisted in, lead to his or her demise. This fact, together with failures so numerous as hardly to be explained away even to the most gullible, is why magic is so much less believed in than religion. "God helps those who help themselves," "Trust in God but keep your powder dry" are bound to be maxims in every theology.

It is a biologically advantageous trait to keep on with efforts to preserve oneself even when in a rational appreciation of the situation the chance of prevailing is very slight. For it cannot lead to *fewer* survivals and subsequent begettings, and must in some instances endow the believer with the extra fillip of confidence that makes him win out. Hence, a tendency to have high beliefs in gods, to be religious, must be favored. From the evolutionary standpoint, *edification is to high beliefs as truth is to low beliefs.*

Optimism is not the only edifying effect of high beliefs. Of even greater import is their usefulness for strengthening social bonds and facilitating cooperative behavior. It seems to be a law throughout the animal kingdom that an individual acts deliberately only with a view toward what it believes will preserve itself and further its individual interest. Since (roughly) all low beliefs are true, this means that animals with no high beliefs will not choose to do things that are in the interest of the herd but have a deleterious effect on the agent: in other words, such animals will not deliberately sacrifice themselves. (Emphasis on "deliberately;" there may be such a thing as instinctive altruism.) High beliefs, on the other hand, need not be true; consequently an individual holding a high belief according to which it is in his or her interest to do something that in fact is opposed to self-interest, but is socially valuable, may be motivated to do it. Loyalty, public spirit, even a moderate degree of habitual altruism become possible when agents come to believe that they "lay treasures up in heaven" by so acting or that they risk divine displeasure and punishment by shirking. Thus high beliefs can help to solve the "prisoners' dilemmas" that constantly beset groups.

High beliefs, moreover, engage the emotions at the deepest levels. They elevate the believer above the humdrum and endow his or her existence with cosmic significance, as also they do that of the family and the community, on whom indeed they confer identity. They define values. They are versified, set to music, chiseled in marble. Important personages, garbed in gorgeous vestments, are appointed to enunciate them inside magnificent edifices erected expressly for that purpose. Higher education, learning that goes beyond the merely utilitarian, may consist mainly, sometimes wholly, in memorizing and commenting on them. They are the community's most precious possessions, to be guarded and preserved uncorrupted at all costs.

We—you and I, anthropologists as it were, viewing from outside—can separate the beliefs current in some tribe into two classes: low (tested) and high (based on imagination, not yet jeopardized by being put to experiential test). This distinction, however, is not made *within* the tribe. To the tribesperson, a belief is a belief—an expression of the truth about how things are and are going to be—no matter how it got established. Even in a simple society, most people will acquire most of their low beliefs by being told; that is, these social low beliefs will be personal high beliefs. Hence, there will be no obvious ground on which the individual can distinguish between beliefs based on somebody's experience and on nobody's. And suspense of judgment, doubt, and unbelief contrary to social pressure are rare in every society, no matter how sophisticated. (Including, of course, our own; the proliferation of "Question Authority" bumper stickers in the 1960s was no counterexample. The orthodoxies of the people inside the cars were even more rigid than those of the un-with-it multitudes outside.)

AN INVISIBLE MEMBRANE

In Eden, high beliefs and low coexist and do not interfere with each other; they are, as it were, kept separated by an invisible membrane, which insulates high beliefs from being jeopardized. It is invisible because, as we have seen, most social low beliefs are individual high beliefs. And inasmuch as social high beliefs are mostly explanatory, expressed in terms of the plausible notions and categories derived from everybody's experience—desire, will, good, evil, goal—they seem quite "natural" and there is no motive to question them.

Once established, high beliefs possess their own inertial mass: as there is little or no ground for challenging them, and so many interests both conscious and unconscious are served by their being kept viable, they tend to persist indefinitely. After all, they enshrine the values of the society: not only do they make sense of the world, provide the framework of explanation, and account for why things are as they are; they define what is worthwhile, what needs protection, what must be furthered, and what cannot be allowed. As such they are instinctively protected. In every society the most assiduously instilled of high beliefs is that of the wickedness of questioning high beliefs.

So all is for the best in the little world of the hunter-gatherer band. In the view of the tribesmen, all of their beliefs are true, they all describe the world as it really is. *We* (anthropologists) know that the tribesmen are mistaken: only their low beliefs are true, their high beliefs are literally false. But the mistake is one the consequences of which are, on the whole, beneficial. All the beliefs held within the tribe, low and high alike, are *right*, where "right" means what furthers coping—which comes to "true" when applied to a frequently jeopardized belief, one that every day rubs up against the reality that the tribesman has to cope with, and "edifying" otherwise. And what is true need not be edifying, nor must what is edifying be true—it need only be *believed*, that is, assented to, held to be true.

This happy modus vivendi persisted for a million years or more. With the invention of agriculture ten or twelve thousand years ago, leading to a sedentary way of life, enormous increases in the size of social units, and invention of the State, the importance of high belief systems mounted concomitantly. For high beliefs are social glue, indispensable for keeping large populations together that are not united by the natural bonds of family and tribe, nor even by common acquaintance. The need was all the greater inasmuch as great disparities in wealth and power, unknown in the band of forty, exacerbated envy, a divisive sentiment.

There are many reasons why the "membrane" is "invisible"—why people within a belief system do not consciously divide their beliefs into high and low with different bases, but take them all as equally true, rooted in experience. The experience, that is, of the *prophet*—the originator of a high belief—which is deemed *more* veridical than ordinary sense experience, with the high indeed as *more certainly* true than the low. Here are some:

1. The inertia of belief. A high belief gets established by contagion. Every high belief must begin in some individual, but private high beliefs are precarious and inconsequential unless they snowball to become "what everybody believes," whereupon they are just like low beliefs, at any rate within the band. Having attained this status they are nearly invulnerable.

2. All persons except the "prophet," the originator of a high belief, acquire the belief by being taught, which is the same way they acquire most of their low beliefs.

3. High beliefs "explain" low beliefs. That is, they make sense of the facts that are the content of the low beliefs. Hence, they are not thought of as separate from those facts. If the ordinary term for epilepsy is "sacred disease," every seizure is regarded as observational confirmation of divine activity.

4. Many high beliefs deal with matters that logically cannot enter into mundane human experience, hence cannot be disconfirmed by it—for example, the fate of the soul after death. And before writing and record keeping, the invulnerability of beliefs about the past, more remote than two or three generations, to experiential checks was almost as absolute.

5. Non-supernatural persons—you and I and the people of our mundane dealings—often behave erratically and unpredictably. So failures of supernatural personages, who after all are modeled on us, to act as expected—for example, not answering reasonable prayers—can easily be discounted; these beings "move in mysterious ways." Answered prayers, of course, count as confirmations.

6. Pseudo-confirmations: oftener than not, beliefs are not about what starkly will or will not happen, but about what is more or less likely to happen. If the weather bureau announces that there is a 30 percent chance of rain tomorrow, and tomorrow turns out to be sunny, has the weather bureau been refuted? Questions such as these, difficult to answer otherwise than arbitrarily, play into the hands of the high believer. He can pray to Zeus for rain, and if it rains, claim it as confirmation of the existence of Zeus. If it does not rain, well, Zeus, being only (super)human, evidently did not *want* to oblige. Beliefs about Zeus, he can hold, are not essentially different from beliefs about bureaucrats to whom one sends petitions and bribes.

7. Probability judgments are a fertile source of superstitions. Disasters to nations and individuals are of such frequent occurrence that the appearance of a comet or black cat can almost always plausibly be held to foretell one. As high beliefs explain things, making use of the human activity model, which is irregular and uncertain in so many of its oper-

ations, the intimacy of the high and the low—the low explained by the high—is thus always reinforced, never weakened, no matter what happens in fact. How can there be any gap between the worlds of the high and the low, when patently it was the prayer to Vulcan that reduced the breakage rate from 17 to a mere 9 percent in the latest firing of the kiln?

8. Imagination makes possible false *low* beliefs, as long as they are relatively harmless: all sorts of superstitions—for example, celery is an aphrodisiac—may persist notwithstanding frequent disconfirmations.

9. It is possible simply to refuse to recognize a conflict between a high belief and a low belief that contradicts it. Logic is low; logical principles are neither innate in the human mind nor learned otherwise than from experience; and in the realm of high beliefs, there is no experience. So not only can conflicting high beliefs both be accepted, but a low belief and a high belief logically incompatible with it can be simultaneously maintained.

Nevertheless, from time to time even in Eden untoward events will happen. Somebody will climb Olympus and find no palaces, only more rocks. Some clod will stumble into the Chief's shadow and emerge unharmed. Must not the high beliefs then collapse? They have rubbed up against reality, and they have worn through; or in terms of the other metaphor, the invisible membrane has been punctured. What then?

The arrangement whereby high and low beliefs are kept separate is so convenient that we should expect to find mechanisms in place to repair such breaches when they occur. And we do find them.

Suppression of the discovery is the most straightforward and is nearly always resorted to at the outset. The bean-spiller is dealt with by social pressures including main force. Olympus is fenced off; mountain climbing is made a criminal offense; people who purport to have been to the top are stigmatized as liars or madmen. The survivor of the Chief's shadow is done away with, along with the witnesses to the debacle.

All is not lost even if these expedients fail. Recourse can still be had to allegorical reinterpretation. Saying that the gods dwell atop Olympus was merely a way of declaring, in language the people could understand, that divinity dwells *on high*—which is not a location at all, but a state transcending space and time. The Chief's shadow does not bring about the death of the body, but inflicts an even more terrible fate: a *spiritual* death.

To sum up the situation of beliefs in the "natural" human social unit, the hunter-gatherer band:

Beliefs are, of course, inculcated and transmitted culturally, not genetically. Nevertheless, tendencies to hold beliefs of certain kinds are built into the genes and are therefore subject to evolutionary constraints. Our genes tell us to believe that fire burns, that objects as they recede from us subtend smaller visual angles, and that some of the noises made by other people convey information. Which is not to deny that individuals' beliefs to these effects must be activated in experience.

Evolution sees to it that the structure of an organism is *right* for its niche: that is, that enough individuals of the kind are successful enough in coping with their circumstances to live long enough to reproduce. This applies to their belief structures also, since those structures are the main determinants of their action, their coping. Consequently it is misguided to complain of a certain tribe that its beliefs are a morass of ignorance and superstition. If the tribe is maintaining its numbers, it is doing something right, and so, because action depends on beliefs, its beliefs are right too—though they may not all be *true*.

The low beliefs of the tribe are right in the only way a low belief can be right, namely, by being true. The high beliefs on the other hand are right because they promote individual efficiency, tribal cohesion, and other social virtues, or at least do not derogate from them. They need to be *edifying*. The particular contents of the beliefs may vary widely from tribe to tribe while still being right. Indeed, their distinctiveness is often an important component of their rightness, inasmuch as possessing beliefs that other people do not subscribe to may serve to define a coherent group and unite it against its competitors—a cultural analogue of distinctive markings and plumages.

Thus taking note of the existence and function of high beliefs—uniquely human possessions—explains how it can be the case both that the human species is a biological success, and that it has a virtual monopoly on false beliefs. In Eden, false beliefs can still be right beliefs if they are high beliefs. And although we do not live in Eden any longer, that is the kind of life that we are still adapted to. This is the answer to our first question.

SUMMARY

About 99 percent of human evolution has gone on while social units were hunter-gatherer bands of about forty adults, which is therefore the "natural" human milieu. We call it *Eden*. The low beliefs of all bands will be much the same, and true. But different bands will have diverse sets of high beliefs, untested for various reasons ranging from logical impossibility to expense.

High beliefs, however, will tend to be *edifying*: promoting optimism and social cohesion. They will *make sense* of low beliefs, which of themselves are mostly unconnected. Making sense must be teleological, that is, explaining what happens on the model of human activity in terms of good, evil, purpose, goal, love, hate, and such like emotions, ascribed to unseen but person-like beings of great power: spirits, gods. They will be conceived as approachable and subject to being influenced by sacrifices and prayers. By these means high beliefs offer *hope*. They will be the most precious possessions of the band and carefully guarded against doubt and disbelief.

In Eden an *invisible membrane* keeps high and low beliefs from getting in each others' way. If a high belief perchance gets rubbed up against reality and is in danger of disconfirmation, there will be mechanisms at hand to keep the membrane intact, from elimination of the disconfirmers to reinterpretation of the beliefs in a more spiritual (and less testable) sense.

In sum, evolution works to assure that the beliefs of a community are *right*: low beliefs true, high beliefs edifying. This is the answer to the question of how it can be the case that humankind is both the most biologically successful species and also practically the only one entertaining false beliefs.

CHAPTER 7

cho

Babylon

Organic evolution on the earth is over. It was a process that went on for three billion years, more or less, and resulted in us and all the other creatures. Then, ten or twelve thousand years ago, our ancestors invented agriculture, which changed everything fundamentally—not only human life but all life. In doing so we altered our role in relation to the earth from the essentially passive one of receiving the nourishment that she produced on her own, so to speak, for the *active* one of interfering, using planting sticks, plows, and whatnot, to turn her operations in the direction of producing, for our benefit, particular products that we desired in greater quantities than she would supply left to herself. Since that first planting—a period of maybe a third of 1 percent of human evolutionary time—our activity has so altered the universal milieu that what survives and what becomes extinct no longer is determined by what advantages or disadvantages accrue in the genotypes in relation to respective prey and predators, but on how the creature in question fits into the human scheme of things. The cetacean brain is as complex as ours and evolved through the same laborious and painful, long drawn-out process. In two centuries cetaceans have been put in danger of extinction because people first coveted their blubber to fuel their lamps, later because they developed a taste for tuna sandwiches and casseroles. Now, perhaps, a happy ending: there is a possibility that in the nick of time they will be spared on account of the sentimentality characteristic of our species.

But from the human point of view, the fact of most interest is that *human* evolution has stopped. The human race is not to the swift any longer (if it ever was). What sorts of people, if any, will be around ten or a thousand generations from now, depends not on what traits, exercised in free competition, prove useful for survival, but on what decisions are made about who is to be allowed to do what and to have what. This does not mean that our immediate and remote descendants will have the traits we hope they will have, and that we consciously try to promote—there has never been a large-scale human breeding program, and perhaps there never will be; but it does mean that they will be as they will be not on account of the working of blind impersonal forces, but on forces that are personal, though perhaps equally blind or even blinder. It is not merely God who is dead; so is Nature.

If the human species were a year old, all but the last day would have been spent as hunter-gatherers. The traits evolved in that mode of life are not necessarily what conduce most to flourishing in civilization; and because the civilizing process has gone so rapidly and shows no sign of slowing down, there has not been time, nor will there be, for evolutionary forces to adapt us as a species to living in groups of millions. We have to do it ourselves, if it is to be done at all.

The concern of this book is not to ponder the general problem thus created, but only one of its aspects, that of the human propensity to believe, and especially to have high beliefs. If the central argument is correct, natural processes have built into the human germ plasm strong tendencies to have true low beliefs and edifying high beliefs. These are what the hunter-gatherer group needs and they are what it consequently has. Man has true low beliefs because only with them can he have a chance, out in the wild, of coping with the world. But because he has language, which makes it possible to form beliefs in another way besides that of direct experience, he can have high beliefs also: beliefs not directly tested in experience, but which nevertheless can affect conduct by being part of the mind-set of the believer.

There is no need for such beliefs to be true, nor are they. But they are *held* to be true—they would not be beliefs if they were not; and doing so is advantageous or at least harmless both to the individual and to his society, when that society is the hunter-gatherer band of forty.

The transformation of human society from an aggregate of hunter-gatherer bands, the way of living that we have called Eden, to Babylon—civilization, citification—was not an evolutionary process at all but a course of development resulting from technological innovation, the kind of human transformation with which we grow more and more familiar in one sense, bewildered by in another, in our own day. This is not to say that it was planned. But certain bands must have decided consciously to engage in cultivation of foodstuffs—perhaps because, according to a plausible theory, having learned to make beer, they found the supply of wild grain for fermentation insufficient to assuage the thirst that quickly developed. At length abandoning nomadism, they had to pay attention to such matters as water supply, which in the Fertile Crescent led to artificial irrigation works requiring cooperative effort from large gangs. Bands did not decide to coalesce into cities, but they decided to do things that could not but lead to cities.

Hunter-gatherer bands *could* exist without religion, that is, without legends of gods. This is clear from the fact that the hunter-gatherer band antedates language (which could only develop in a social situation), hence, imagination and religion. And anthropologists have encountered bands whose creation legends, for instance, seem impersonal. Edifying legends, we have argued, are a survival plus, and very natural; but not absolutely indispensable.

It is otherwise in Babylon. Cooperative unity of hundreds of people, let alone millions, is not natural; its development and maintenance is the fundamental problem in civilization. The virtually universal presence of religion in the hunter-gatherer bands that initiated civilization was, as it were, providential: the social glue was already there. More exactly stated, the propensity to form and propagate edifying high beliefs is what, in the human germ plasm, makes civilization possible.

If three or four bands, each with its own high belief structure, coalesce into one proto-city, which of the three or four religions prevails? Like the question of which language will dominate, this is a matter that can be resolved in many ways. One of them may be imposed by force, or numerical predominance, or greater psychological attractiveness, or accommodation—after all, each religion is apt to have a head god, and a goddess especially of women, and so on, and these can be conflated (as Venus = Aphrodite, Ares = Mars, etc.); or each religion can go on, for a while, as the special cult of a particular deme, with one cult, even a new one produced for the occasion, as a super-pantheon for the whole city; this last was exemplified in the coexistence of Olympian gods for the whole city, and lares and penates for the clans, in our ancient world. All of them work, eventually; all of them achieve the desired end of uniting the citizens in a communal bond transcending the natural band.

The expanded physical resources of civilization further the process. Instead of one shaman, on terms of familiarity with the thirty-nine other members of the band, making difficult the maintenance of a mystique of awe, there can be a special class of *priests* with special lore, privileges, and rituals quarantining them from vulgar scrutiny; special (and sumptuous) buildings set aside for their exclusive use; and arrogation to them, exclusively, of the function of communion with the gods—a privilege bestowing enormous authority within the city upon its holders.

In the hunter-gatherer band, the shaman is one man (or woman), the leader of the marauding party almost always another. The civilized priest is the descendant of the former, the king or emperor of the latter. Maintenance of civil order requires that these two powers operate in close harmony. The most direct means of ensuring cooperation is to combine the two offices: hence priest-kings. On the other hand, priesthood and military leadership are such different types of occupation that one person is seldom good at both, hence, in fact if not in constitutional theory, two different individuals will usually exercise supervision over the functions. But the ruler will always enjoy a special religious status, up to and including divinity. Separation of church and state was never dreamed of before an hour ago, as it were; the question of its ultimate feasibility is by no means yet settled.

In the area with which we are primarily concerned, the Mediterranean basin plus the "Near East" as far as India, there was a major difference in the ways in which civilization developed. In the eastern part and in Egypt, the tendency was for consolidation, by conquest, into large empires, culminating in the Persian, approaching a million square miles; while in the West, political units at this time hardly ever exceeded the modest size of the island of Crete; mostly they were individual cities, sometimes joining in temporary and precarious alliances. This differentiation probably was due mainly to the necessity of artificial irrigation in the East, while the West, except for Egypt, was dry farmed. It had a profound effect on their respective conceptions of the relations between and modi operandi of their respective gods, as we shall see; and on the possibility of individual thought occurring and gaining a hearing.

SUMMARY

Organic Darwinian evolution is over, having been brought to an end by the vast explosion of technology that enables people to determine which species shall survive and which perish. This includes the evolution of people themselves.

Though we live now in groups of many millions, the evolutionary stage at which we are stuck is that of the forty-member hunter-gatherer band. Edifying high beliefs, providentially there already when agriculture and civilization began, took on vastly enhanced importance as the social glue making cities and states and empires possible. And instead of the individual shaman of the band, civilization required and produced a class of professional priests, whose prestige and livelihood was bound up with the preservation of high beliefs.

Political units in Egypt and the East were huge, whereas in the West they seldom exceeded the bounds of individual cities. This difference had a profound effect on the conceptions of the universe current in them.

PART II

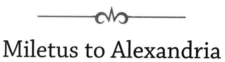

Miletus to Alexandria

CHAPTER 8

✿

Miletus: The Invention of Science

Nevertheless, systems of high beliefs, like all things human except low beliefs, are mortal. They come to their ends, however, only to be replaced by other high belief systems—with a single exception, to which we now turn.

Spinoza, with his usual remarkable insight, observed in the Appendix to Part I of his *Ethics* that

> truth [about the nature of things] might have evaded mankind forever had not Mathematics, which is concerned not with ends but only with the essences and properties of figures, revealed to men a different standard of truth.

Whether or not he knew it, Spinoza was referring to one man's achievement at a certain definite time and place: that of Thales, son of a Greek mother and Phoenician father, in the city of Miletus near the beginning of the sixth century B.C. Only once in history has this happened; wherever science, whether natural or mathematical, exists it is descended from Thales' innovation.

Miletus was the largest and most important of the twelve Greek cities of Ionia, along the coast of what is now Turkey. All these cities lived by manufacture and trade, their hinterlands being under the control of eastern empires. They were the distant heirs of the Cretan or "Minoan" civilization of the second millennium B.C., which collapsed after the Trojan War from unknown causes, ushering in the four-century period of the "Greek Dark Ages." The cultural linkage between the two periods was tenuous, mainly the institutionalized recital of long poems about incidents in and around the Trojan War. When about 700 B.C. the Greeks made themselves literate by borrowing the alphabet from their Phoenician (Semitic) neighbors, their first productions were the *Iliad* and the *Odyssey*, traditionally attributed to "Homer," who probably was two (or more) persons writing down compositions with long previous histories as oral recitations. These poems were the first works of European literature and are still its greatest. It is the amazing characteristic of the Ionian Greeks to have been first and greatest not only in epic poetry but in a multitude of other fields: history (Herodotus), lyric poetry (Sappho), logic, biology, literary criticism, ethics (all Aristotle), and—above all, for it absolutely changed the world—science: Thales' invention. This is a big claim to make for a man who wrote no book and about whose teaching all we know is contained in a few brief and garbled reports made centuries after his death. But these broken tiles outline a mosaic picture of a man of more than Einsteinian stature:

- He went to Egypt and learned the lore of the priests. {D/K 11A1; from Diogenes Laertius}
- He invented geometrical proof. The theorems that vertical angles are equal, that the base angles of an isosceles triangle are equal, and that the diameter of the circle bisects it, are ascribed to him. {D/K 11A20; from Proclus}
- He held that the All is One. {D/K 11, passim}
- Moreover, the stuff that basically everything is composed of is water. {bid.}
- He held that the earth floats on water. {D/K 11A14; from Aristotle}
- He claimed that the stone of Magnesia has a *psyche* because it can move iron. {D/K 11A22; from Aristotle}
- He predicted "the very year" (585 B.C.) in which a solar eclipse would occur in his part of the world. {Herodotus I 74}
- He said that all things are full of gods. {D/K 11A22; from Aristotle}

Perhaps this does not sound like all that much. But consider:

Egypt was a land both of superstitions and of advanced technology importantly based on knowledge of various *facts* about numbers and areas: true low beliefs. But typically of low beliefs they were conceived in isolation from each other; no priest had the idea of how they might *hang together*, be shown to form a unified body of knowledge by *proving* them, that is, deriving them from some of the simpler ones whose truth was intuitively obvious by using logical principles (also low beliefs!).

Let us look at one of the theorems of geometry that Thales is said to have proved: that vertical angles are equal.

The following may or may not be the proof that he produced. It does not matter; for any proof will exhibit the features to which we shall call attention.

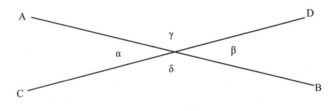

Given: AB and CD are straight lines intersecting in the point O and forming the two pairs of vertical angles, α/β and γ/δ.

To prove: $\alpha = \beta$ and $\gamma = \delta$.

Proof:

1. $\alpha + \gamma =$ a straight angle. Reason: definition of straight angle as angle whose sides lie on the same straight line.
2. $\gamma + \beta =$ a straight angle. Same reason.
3. Therefore $\alpha + \gamma = \gamma + \beta$. Reason: All straight angles are equal (axiom).
4. Subtracting γ from both sides of the equation, $\alpha = \beta$. Reason: If equals are subtracted from equals the remainders are equal (axiom).
5. By similar reasoning, $\gamma = \delta$,

Q.E.D.

Perhaps the most striking—and apparently disappointing—feature of both theorem and proof is their utter obviousness. Anyone who has drawn an X can *see*, right off, that the top and bottom angles are equal and so are the other two. Probably there were Philistines among the merchants calling at Miletus; if so, they may well have ridiculed Thales for wasting time in proving what everybody knew already.

The Philistines would have missed the point. Although everybody knows that vertical angles *are* equal, not everybody knows *why* they *must* be equal. Thales had shown that to be equal *follows from* what it is to constitute a pair of vertical angles, which for this reason are *necessarily*—*never-ceasingly*—equal. Perhaps this knowledge was in some way implicit in anyone who "saw" that vertical angles are equal; all the same, it was an advance to make this knowledge explicit. And the proof *articulated*, began the systematization of, what had before been separate items of the lore of measurement, engendering a cumulative process. When in proving that the base angles of an isosceles triangle are equal—another obvious theorem—he needed to assert that another pair of angles were equal, he could simply give as his reason, that they were vertical angles, whose equality had been "previously proved." And similarly in the proof of the not quite so obvious theorem that any angle inscribed in a semicircle is a right angle. With such helps, a couple of generations later Pythagoras was able to prove his famous theorem, which is not obvious at all. Discovery of the technique of proof is the zero point from which science proceeds, analogous in the intellectual sphere to the first appearance on earth of multicellular living things.

Thales was the only begetter of the natural sciences too, and in similar fashion: showing how, following up on the intuition that things *do* hang together, that the All is One, low beliefs previously unsuspected of having any connection nevertheless could be brought together and explained in a unified theory without recourse to high-belief type entities.

We have listed above four pronouncements ascribed to Thales:

- That all things come from water;
- That the earth floats on water;
- That the stone of Magnesia has a *psyche*; and
- That all things are full of gods.

How can these, especially the last, add up to a claim that Thales was the first natural scientist?

Let us begin with the third and fourth. "Psyche," which basically means "breath," at this point in Greek thought signified the life principle (whatever it is). So Thales seems to have attributed life to a stone. But this is not so primitively animistic as it looks. The *power of motion*, Thales is saying, not some mystical afflatus, is what distinguishes the animate from the inanimate. The magnet exhibits this power in a simple way. Likewise for the fourth thesis, that all things are full of gods: the Greeks, like everybody else, supposed that matter is inherently lumpish and inert and must be moved from outside, ultimately by "gods," supernatural forces. In both statements, Thales is making the revolutionary pronouncement that to the contrary, motion is of the nature of matter, inherent in it—which means that what it *does* can be adequately explained in terms of what it *is*, just as the equality of vertical angles can be explained in terms of what they are.

As for the notorious pronouncement about water: let us first note that Thales did not say that "Everything *is* water," as the usual summary has it; he said something that Aristotle paraphrased as "Water is the originating principle (*arché*) of all things"—everything *comes from* water. But he did mean that in some sense, there is just one basic stuff, and that stuff is what we ordinarily identify as water. The thought that there is one basic stuff is more significant than its identification. Thales is implying that the world is a unity, in which everything is connected with everything else; moreover, the respect in which it is a unity is one that ordinary human cognition is capable of grasping. Water is familiar to everybody; it is not the special preserve of shamanistic insight.

If Thales had an account of the manner in which everything "comes out of" water, it has been lost. But we can infer what his crucial insight was: things come from water because it is the nature of water to produce them—they are not fished out of it by supernatural beings. His model, it seems, was taken from the troublesomely continuous silting up of the harbors of Miletus.

The picture of the earth floating on water came perhaps from regional mythology. The detail about earthquakes shows Thales' unifying mind at work. First, it is an explanation of the new type: something extraordinary, puzzling, and catastrophic is explained as a natural occurrence, analogous to a commonly observed phenomenon, rather than as the intervention of a personal force (Poseidon). Second, it is a quasi-deduction from the general theory: everything comes from water, the earth floats on water, and, as a floating body, is subject to what ordinarily happens to floating bodies, namely, it tosses about when the water gets rough. A poor explanation, *we* may say, connected only tenuously to observation, and vulnerable to many objections. Yes, but it is of the right *type*. The Milesians might have objected to it on other grounds: you can, in principle, do something to appease Poseidon; you can do nothing to keep subterranean water from being stirred up. To accept this type of account would then deprive them of a kind of hope—which deficiency, as we shall see, became a serious emotional check on acceptance of science.

World-origin myths around the eastern Mediterranean were virtually unanimous in holding that the earth came out of an original expanse of water. There was some factual (low belief) support for this contention in the ongoing process of silting in the river deltas. But all the myths had this form: The earth-containing water just sat there until gods *distinct from and outside* the water used their power on this inert stuff to separate out dry land. They engaged in this activity *for the purpose of* . . . whatever. This was the high-belief component of the myth.

Thales did not dispute the low belief part of the account. What he did was eliminate the high belief: not by simply subtracting it—there would then be no explanation—but by incorporating its essential feature, which is power to bring about change, into the stuff to be transformed. "All things are full of gods," he said. But his gods were *inside* the stuff and not distinct from it; they were what later came to be called its Nature or Essence (Being). And these "gods" were stripped of their anthropoid features, most notably their desires and purposes.

We are now in a position to describe that unique event as the creation of the first *worldview based on low beliefs*. Thales produced the first *scientific* Grand Unified Theory of Everything. It differs from its mythological predecessors clearly and sharply in being free of high-belief type elements and in possessing three features that are essential to science:

1. *Unity, Monism, Reductionism.* "The All is One"; there is an underlying unity in the apparent diversity of things. There is only one kind of reality.
2. *Essentialism, Naturalism, Immanence*: The All contains within itself its driving force; its changes are effects of internal causes.
3. *Rationalism, Necessitarianism, Logos*: The Principle of Sufficient Reason—"There is a sufficient reason why everything that is, is so and not otherwise"—holds universally; there are no "brute" (uncaused, inexplicable) facts: "Nothing occurs at random, but everything for a reason (*logos*) and by necessity," as a later Milesian sage (Leucippus) put it.

Thales conceived, moreover, that like proof in geometry, the water theory afforded a basis for unifying low beliefs hitherto separate into a complex whole. Earthquakes and facts as distant in conception from them as that the seeds and nourishment of all living things are moist, achieve connection in its light. The remark about the magnetic *psyche* was a demystification of the life principle, perhaps overshooting the mark to find a connection of organic and inorganic.

We do not know what is behind Herodotus's obviously garbled story of Thales and the eclipse of 585 B.C. (so convenient for dating these matters), but it does illustrate Thales' great interest in astronomy, as does the story of his falling into a well when, one night out walking and gazing at the stars, he failed to look where he was going. {See Plato *Theaetetus* 174.} Indeed, one "book" is plausibly attributed to him: a *Nautical Star Guide*, containing the important advice to steer by the constellation Ursa Minor (containing the pole star) rather than Ursa Major.

Two further points need to be made about the significance of Thales' introduction of proof. The first is that *mathematical and logical beliefs are low beliefs.*

Considering the awe in which philosophers have held logic and mathematics, it may come as a surprise, almost seem a blasphemy, to have them described as "low beliefs," along with the cat is on the mat, wet clothes will dry when hung out in sunlight, and the seasons follow each other in a regular sequence. However, they clearly fit the definition of true low belief: they are in fact continually tested by rubbing up against the reality that they are about, and when so tested they promote coping with that reality, never resulting in discomfiture to the tester (if he or she gets them right). Further, they are learned by experience, a combination of being told and trying them out for oneself. This in itself says nothing about what *makes* them true.

The second point is an extension of the definition of low belief: henceforth we shall include among low beliefs, belief in any statement necessarily following from a statement that describes a true low belief; or less pedantically, allowing (as ordinary language does) beliefs to have implications, any belief that necessarily follows from a true low belief is itself a true low belief. That the implications of true low beliefs are true follows from the principle, itself a low belief, that logical implication is truth-preserving. That those implications are also low beliefs follows from the insight—another low belief—at the basis of proof, that whatever is evidence for the axioms is evidence for the theorems also. That indeed is just what a proof is: a method of concatenating low beliefs into a system. This is a point of crucial importance; for as we have noted, before the development of geometry as a science, low beliefs were "loose and separate," isolated entities from which little or nothing beyond could be inferred. They could not constitute something so broad in its scope as a worldview.

Proofs explain—answer the question "Why is it this way?"—by showing that it *must* be this way, there is no real alternative, it is thus of necessity. This is entirely unlike the previous mode of explanation, "Because such-and-such a personage (natural or supernatural) willed it."

Thales' way of looking at the world, dispensing in principle with high beliefs, at last (as Spinoza noted) offered alternatives to the human-action model for explanation: instead of good, evil, will, purpose, love, and hate—all unobservable in nature at large—the new impersonal conception of causation was in terms of observables. It was in head-on conflict with the worldview that over millions of years had become literally instinctive—and still is, for an overwhelming majority of the earth's inhabitants at the present day.

Some features of the way people before Thales looked at the world (and after him, up to and including the present day, to a large extent) are these:

There is a foreground and a background. The background is all the commonplace features of the world, what everybody takes for granted, what universally held low beliefs are about, what is always the same, what is *invariant*. Everybody knows about the succession of the seasons, the moon's phases, the migration of birds. These plants are tasty and nourishing, those are poisonous. Your hunting will be more successful if you manage to keep downwind of the animal you are stalking. You can't make fire with wet tinder. Eggs solidify when boiled. . . . These are specimen examples of *how the world is*, features seldom noticed explicitly, and even when noticed, not thought to call for comment or explanation.

The foreground consists of the interesting and *variant* features, *news*, those things and happenings that are of singular or rare occurrence, often affecting our weal or woe: comets, plagues, droughts, earthquakes, extraordinarily favorable or unfavorable growing conditions, wars, the favor or enmity of powerful people. These are thought of as demanding explanation—more in order to control them than to satisfy curiosity. And the way they are explained is in terms of the model of human purposive activity. This is so because all of us, being always involved in such activity, know what it is and how it works (or at any rate we think we do). We are all the time trying to advance our own interests and those of our allies, and trying to thwart the designs of those opposed to us. And other people, after all, are always doing strange things—sometimes nice, sometimes awful. It is reasonable, therefore, for us to extrapolate this scheme and to suppose that anything that happens and has a serious effect on us is the outcome of some such personal activity. In other words, the foreground features of the world are all the results of personal agency. The categories of explanation, of these foreground features, are such as good, evil, will, purpose, friendship, and enmity. These are the concepts in terms of which we will construct our account of the world, simply because we have no others. Or rather, although there is another schema at hand—that of the way things go, the operations of the background entities—those are not thought of as being explanatory at all.

But the more we learn about the background—become acquainted with the constant properties of things and how they affect one another when left to themselves—the more materials are at our disposal for an alternative general account. We can now see that in the worldview of Thales, *the background has taken over the foreground.* In his thought the progression from water to dry land is no longer a rearrangement of things by personalities; it is the working out of the inherent, invariant nature of water. The earthquake is no longer the shaking, from outside, of the land by another indignant or mischievous megaperson; it is the manifestation—on a large scale, to be sure—of the ordinary behavior of a floating object. Magnetic attraction is not magic (subjection to an exterior will); it is a property—a strange one—of a certain kind of stone, perhaps the simplest kind of vital power.

In nature it is processes, not things, that are invariant. Although human effort cannot change necessity, it can, by understanding it, take advantage of it. Perhaps this is the point of another garbled story about Thales: that he became rich by knowing more than other people did about the workings of the weather, thereby being able to predict it and profit in the futures market for olive oil. One can make rational predictions only on the basis of natural invariants. Thus, viewing the world as the interplay of necessities is not necessarily a philosophy of resignation.

Indeed, Thales seems to have rethought the character of will-guided processes. Some of them, building being the most notable, do seem to involve only the imposition from outside of intelligent forces on inert materials: the inherent properties of lumber, stone, and clay play no active part in construction. In manufacturing processes like pottery making and ore smelting, however, the potter or metallurgist must rely on the inherent properties of fire, ore, and glaze to produce the desired result. He can only bring them together, then wait for their inner invariant natures to manifest themselves. It is this kind of process that furnishes the model for the new Ionian natural science.

WHY DID THALES "LEAVE MARDUK OUT"?

The cosmogonies of the Babylonians and Hebrews depict the earth, "dry land," as arising from an initial condition of nothing but water. The Babylonian god Marduk effected this by dredging up mud and building an island atop the corpse of the slain goddess Tiamat. In Israel, Yahweh, moving upon the face of the deep, utters the command: "Let the dry land appear," whereupon it does. No doubt the reason for consensus on a watery origin of things lies in the fact that the nucleus of the Babylonian civilization was the flood plain of great rivers, where inundations, "watery chaos," recurred, requiring extensive rebuilding after their recessions, with much hard labor by the fellahin under the direction of exalted rulers. (Palestine is not subject to flooding; but the cosmogony of *Genesis* was derived from Mesopotamian sources.) The story-telling imagination would naturally model a general account of origins on this familiar experience.

Thales, who lived in that part of the world, must have known of these water cosmogonies, and that fact may partially explain his own adoption of water as the origin of everything. Indeed, Farrington saw only one significant difference between Thales and the author of the Gilgamesh epic, namely, that Thales "left Marduk out," that is to say, gave an impersonal account. {See Farrington ch. 3}

But that difference made all the difference. It is no exaggeration to say that "leaving Marduk out" was the Big Bang from which science emerged. With Marduk in, you have a story, of people-like beings carrying out their wills, achieving their purposes, creating things that seem good to them out of materials that are there, but inert, the passive objects of the persons' exertions. With Marduk out, you have the first general account of things that is not a story. The materials themselves, without being shoved around by personal forces exterior to them, bring about a new condition out of an old one, because it is the nature of these materials to alter according to their internal patterns. This is an entirely novel way of looking at the world.

Why did Thales leave Marduk out? Let us first note a relevant difference, overlooked by Farrington, between Miletus and Mesopotamia. In the latter, flooding was a problem; protecting the dry land from the encroachment of water required a lot of human effort. The reverse was the case in Miletus, where the problem was the rapid silting up of the city's two harbors. (Today the site of ancient Miletus is a swamp far inland.) That meant that for Thales the emergence of dry land from water was a natural process, requiring no human help; on the contrary, men had to work hard to counteract it. Thus the experience underlying the Gilgamesh account made human purposive activity an indispensable factor in the production and maintenance of dry land; but there was no such personal causation underlying Thales' model. Marduk was not "left out" as one might conceivably rewrite Hamlet cutting the part of Polonius; he simply dropped out of the picture. There was nothing left for him to do; the "essences and properties" of water—its indispensability for all life and for all seeds, its ability to exist as solid, liquid, and vapor, its certified-by-experience ability to "produce" silt out of itself: these now did all the work of explanation.

SUMMARY

Science began only once in human history: in the city of Miletus in Ionia, early in the sixth century B.C. And it was the creation of one man, Thales, who invented both mathematical proof and natural science. Proof is the technique of combining hitherto disconnected facts into an integrated whole of statements that are *necessary*, that *could not be otherwise*.

Natural science similarly combines isolated facts (low beliefs) into a theory consisting of beliefs which though not rigidly connected by purely logical bonds are nevertheless *tethered* to each other. This was the first worldview based on low beliefs: a *scientific* Grand Unified Theory of Everything. It, and its successors, exhibited three characteristics: Unity (the All is One), Immanence ("All things are full of gods;" the energy of change is not pushing and pulling from outside but inherent in the things that change), and Reason ("Nothing happens at random but everything for a reason and by necessity"). Instead of good, evil, purpose, love, and hate, its categories of explanation were cause, effect, nature, regularity, necessity.

CHAPTER 9

Anaximander and Anaximenes

But Thales' achievement might have come to naught had he not had
a "pupil and associate" of comparable brilliance, Anaximander. Also
an eminent citizen of Miletus, ten or fifteen years younger than Thales, he
carried on the scientific way of thinking of the great innovator to astounding
lengths. Of like importance, he was the first to be in a position to initiate the
last essential component of scientific procedure: *dialectic*; and he did it.
Unlike a shaman's acolyte, he did not hesitate to subject the Master's ideas
to critical scrutiny, modifying or rejecting them when he found them
wanting.

Thus agreeing that the All is One and that facts can and should be
explained in terms of other facts, he nevertheless averred that transforma-
tions of water are insufficient to explain how the present state of the All
came about, for "if one of the elements [earth, mist, fire, water] were infinite
it would have swallowed up all the rest." {D/K 12A16.} Water dissolves
earth, absorbs mist, and puts out fire. So, he inferred, none of these definite
things can be absolutely basic; all four must emerge from something *apeiron*
(indefinite or infinite or both) from which hot and cold, wet and dry, are
"capable of separating out," on account of its rotation.

Furthermore, Thales' notion of an earth floating on water merely raised
the further question of what supported the water. Anaximander took the
breathtaking step of declaring that nothing holds up the earth; it stays where
it is "on account of its equal distance from everything." {D/K 12A11; from
Hippolytus.} What he meant, and how equidistance could keep the earth
in place, must be explained in terms of Anaximander's picture of the
universe.

The earth, he held, is drum-shaped, three times as wide as thick. We
live on one of the flat surfaces. Taking the diameter of the earth-drum as

the unit, at a distance of twenty-seven units out, a torus (doughnut or inner-tube shaped hoop) of mist filled with fire circles the earth, inclined at an angle (the ecliptic) to the flat surfaces. In this hoop, facing us, is a hole through which the fire shines. This shining hole is what we call the Sun; the rest of the hoop is invisible. (Invisible mist can obscure even bright lights, as we can observe on a foggy night.) Concentric to this hoop but closer to us (18 units) is a similar hoop with a dimmer hole: this is the Moon. Yet closer, nine units, are the stars (presumably more hoops, but the account of just what they are and how they work is lost). Eclipses are temporary stopping-ups of the holes (again, we do not know how or by what).

Anaximander's reasoning in support of this picture has not been preserved, but it seems clear that the hoops idea is his means of ensuring that the universe, with the earth at the center, is really symmetrical although to first inspection it does not appear to be. And if it is symmetrical, then the earth can stay still even without support because there will be no *reason* for it to go any one way rather than another. This is the principle on which scales for weighing things are constructed. More generally, it is the Principle of Sufficient Reason in negative form. But there is more to it than that. The idea presupposes a notion of gravitation, that is, heavy bodies in an *asymmetrical* situation will move toward the heavy side of it. It also does away with the "dome" or "vault" picture of the heavens hitherto taken for granted. There is now an empty *other side* of the earth through which sun, moon, and stars can pass: Anaximander is on the way to a conception of an infinite unenclosed *space*, if indeed he had not already attained it.

Anaximander gave an explanation of how this symmetrical system had come into being. The *apeiron*, he held, is by its nature eternally in rotary motion. It is a whirlpool, in which heavier things sink to the center, lighter move to the periphery. Thus the "elements," earth, water, mist, fire, separated into layers. The fire at the outside heated the mist and water lower down until an explosion—a Big Bang!—occurred, wrapping the fire into the hoops of mist, and evaporating the water into more mist; a process still going on, as shown by the presence of marine fossils on hilltops. Life, including human life, originated in the sea; human beings are descendants of viviparous sharks who came out on land. They must have been very different in the beginning from what they are now, for human babies are helpless for so long that they would all have died before maturing if their simpler progenitors had not evolved societies that could give them the necessary protection.

Much or all of this account was written in Anaximander's book, the first prose treatise in European literature. It may have been a companion text to the first attempt at a world map, which he drew.

Alas, only a single sentence (or sentence fragment) has survived from that book:

> Into that from which things arise, so also is their destruction, as is required; for they pay penalties and make reparation to each other for their injustice, in due time. {D/K 12B1; from Simplicius.}

This is an expression of the idea of *cosmic balance* that permeates Anaximander's system. He is obliged to express it in the vocabulary of crime and punishment, for he has no other; but personal agency has dropped out. The things "make reparation" *to each other*—by their nature, not under the compulsion of a cosmic parole officer.

To be sure, Anaximander's hoops, like the gods of the myths, are postulated entities, but with an all-important difference. They are *tethered* to low beliefs, whereas gods are entirely free-floating. Indeed, they might not have been beliefs at all, but rather *hypotheses* entertained pending verification, as proper scientific methodology demands, though it is perhaps implausible to attribute so hygienic an attitude to this pioneer.

A reader of the present day may be inclined to give Anaximander high marks for imagination, but to be skeptical about what such flights of fancy have to do with sober science, then or now. Well, not all that sober, perhaps; Anaximander at the very beginning of scientific thinking was not restrained by consciousness of how many fine hypotheses have come to grief; but aside from this enthusiasm, Anaximander's thought is easy to recognize as being of the same *type* as that of, say, Kepler and Darwin. Further, the first world-description based on low beliefs *had* to be imaginative. There is no imagination in recitals of low beliefs, which even the beasts have. *Imagination is the gift of high beliefs to science.* Science could begin only after a corpus of low beliefs was fertilized, so to speak, by this product of millions of years of high believing. Nietzsche was right to declare that "for countless eons the intellect brought forth nothing but errors." {Nietzsche 1974, Sec. 110.} However, there was not and could not have been a shortcut around those eons on the way to the eventual emergence of truth.

Anaximander built a grander speculative edifice on shakier foundations of observation than is to the modern taste; but it is a quantitative not a qualitative difference. And let us not forget that both with respect to grand principles and details Anaximander often got it right. His conviction that what is now land had once been sea bottom was based on observation of marine fossils on inland hilltops; his inference about human evolution was drawn not only from the general theory of a watery initial stage, but on the sound thought that animals like us, whose young are so helpless for so long, could not have survived in a "state of nature" before societies had developed, and therefore our remote ancestors must have had different characteristics. It is astounding that a theory of both cosmic and organic evolution, worked out in detail, was produced at the very beginning of scientific thought. But it happened.

We lack information of Anaximander's mathematical accomplishments and interests. We can, however, be sure that he had them, from the occurrence in his cosmic description of definite numerical values, in earth-radii, of the distances of the celestial hoops from the earth. Moreover, the Anaximandrian natural-scientific "must," like the Thalesian, was akin to the mathematical: the Big Bang *had* to happen because the natures of fire, fog, water, and earth are such that when tightly layered, an explosion was inevitable; and there *had* to be a stage of cosmic development when they were tightly layered, because ... The non-ceasing natures dictate the outcomes *of necessity*, they *could not be otherwise*. The three Milesian principles of scientific thought, Monism, Naturalism, Sufficient Reason, are there explicitly and unviolated.

ANAXIMENES

There was yet a third Milesian: Anaximenes, whose theory based on Air (or rather Mist), tried to mediate, Goldilocks-like, between too-definite Water and too-vague Boundless as basic stuff. Not intellectually the equal of Thales and Anaximander, nevertheless he it was whose worldview became the "standard" Ionian account for the succeeding century or so. He too preserved intact the basic requirements of science: Unity, Immanence, and Reason.

In the cosmogony of Anaximenes no suggestion of quasi-moral motivation remains: the cosmic process is reduced to the thickening and thinning of the original stuff, which, according to Anaximenes, is neither water nor "the unbounded" but *aer,* mist or fog. Thinned, it becomes fire; thickened, first water, then earth.

Some historians cast doubt on the status of Milesian thought as genuinely scientific on the ground of its well-nigh complete reliance on a priori insight; natural science, they allege, must found itself on observation and experiment. But surely the originators of the whole enterprise—and unquestionably the Milesians were that—should be cut some slack. Moreover, cosmogony, their principal interest, to this day cannot by its nature involve experiments. More observations are now available to it, but only by use of fantastically expensive apparatus providing data to which theories such as the Big Bang are not related deductively but still only by what we have called *tethering*, a process that has perhaps been refined somewhat.

Anyway, Anaximenes did perform an experiment, from whose debacle a moral may be drawn.

In his Grand Theory of Everything the "elements" earth, water, mist, fire are basically one stuff at different degrees of *thickness*: mist is thickened fire, water thickened mist, earth thickened water. He held also that thickening and cooling are the same process. The experiment (which you, reader, can perform right now) was this: Blow on your hand with mouth held wide open: the "thin" air feels warm. Blow on it hard through pursed lips: the "thick" breath feels cold. This was supposed to confirm the connection of thinning with heat and thickening with cold. Actually the experiment illustrates the peril of experiment in absence of developed theory, for the true explanation of the phenomenon involves the cooling effect of *expansion* of the hard-blown air, directly contrary to Anaximenes' theory; also of evaporation of sweat off the palm. But there was no way that a 6th century B.C. Milesian could have known about these matters. Moral: The interpretation of data is seldom if ever independent of theory.

Finally let us repeat our observation that Thales and Anaximander began a dialectical process of criticism and revision, "conjecture and refutation," that is as much of the essence of scientific activity as it is anathema to high belief. Even though high beliefs do in fact sometimes get modified, the notions of explicit criticism and emendation, or of elements of the system that are more or less probable, is inapplicable to them. Infallibility automatically appertains to the authorities and to their pronouncements. As Wittgenstein pointed out, in the theological language-game it makes no sense to say that "Probably there will be a last judgment." {See Wittgenstein 1966.} The "certainty" of high beliefs betrays their origin in stories, to which the vocabulary of probability is likewise inappropriate. Thales and Anaximander may not personally have welcomed criticism: few of their eminent intellectual descendants do even today; nevertheless, criticism got built into the standard procedure at the beginning, and was explicitly noted as proper and necessary, as early as Xenophanes (in the next generation):

The gods indeed did not reveal everything
to mortals from the beginning; but men search,
and in time they find out better. {D/K 21B18.}

From Thales' day to our own, the advance of the objective, scientific conception of the nature of things, though frequently slowed, sometimes stopped, and once or twice even reversed, has proved inexorable. This is the triumph of low belief.

SUMMARY

Anaximander, Thales' "pupil and successor," wrote a book, the first general account of the world in the new way. The notions of evolution (cosmic and organic), survival of the fittest, postulated entities tethered to low beliefs, the balance of nature, and even gravitation appear in it. It was highly imaginative, but it had to be; mere disconnected low beliefs are boring. Science could appear only after the imagination had been "fertilized," as it were, by eons of high believing.

Anaximander further deserves the credit for introducing into science *dialectic*, criticism and (hopefully) improvement of the views of one's predecessors. Holding that water is too definite in its nature to be the source of everything (as Thales had pronounced), he made the original stuff to be The Boundless, something not particularly wet or dry or hot or cold but "capable of separating out" into these.

The third Milesian, Anaximenes, tried to mediate between the Water of Thales and the Boundless of Anaximander by making Air (or rather Mist) into the basic stuff, with thickening and thinning as the process by which change occurred. This became the "standard" Ionian physics for the succeeding century. He further tried to validate this theory by an experiment, which, however, showed instead the futility of experiment in the absence of antecedent theory.

{See Robert Hahn, *Archaeology and the Origins of Philosophy* (SUNY Press 2010) for a stunning, empirically supported archaeological investigation of what Anaximander further owed to observation of the practices of temple builders in the formation of his cosmic picture.}

CHAPTER 10

✧

Science and Philosophy Come to Italy

In the middle of the sixth century B.C. the Persian Empire took over the Ionian cities, causing great dislocations among the Greek populations. Xenophanes, a citizen of Colophon, about forty miles north of Miletus, fled across the Aegean Sea to southern Italy, the region which together with most of the island of Sicily was called *Magna Graecia*, "Great Greece," by the Romans. A young man of twenty-five, he had "heard" Anaximander and thus brought with him an acquaintance with, and a commitment to, the great Milesian revolution in thought. He lived past the age of ninety, making his living as an itinerant poet, reciting his own verses and those of others in the houses of aristocrats.

He had some scientific interests, notably in geology, in which he supplemented Anaximander's observations of marine fossils on hilltops with the discovery of fossilized leaves lower down. He inferred from these that the earth was not simply drying up, as the Milesian had thought, but was undergoing cycles of wetness and dryness. This made explicit the notion of cosmic time as cyclical, which was implied by the unanimous belief of Greeks that the basic stuff of all things could not have come out of nothing but had eternally existed.

Xenophanes was conscious of the incompatibility of the new thought with the traditional religion of the Olympic gods. He declared that people made the gods in their own image, just as

> if oxen and horses or lions had hands, and could paint with their hands, and produce works of art as men do, horses would paint the forms of the gods like horses, and oxen like oxen, and make their bodies in the image of their several kinds. {D/K 21B15.}

He objected to these deities on moral grounds also, pointing out how Homer and Hesiod depicted them as habitually engaged in activities that would be outrageous if people did them. And he was contemptuous of popular superstitions, making fun of a believer in transmigration of souls who called upon a man to stop beating his dog, for he recognized in the yelps of the beast the voice of a dear departed friend.

Nevertheless, Xenophanes had his own rationalistic conception of a proper and true god—like Anaximander, who called his Boundless "ageless and deathless" (the essential attributes of divinity), and like Thales, who will be recalled to have said that all things are "full of gods."

> One god....neither in form like mortals nor in thought. All of him sees, all of him thinks, all of him hears. He stays forever in the same place, moving not at all; nor is it fitting for him to go about, now here, now there. But without labor he shakes all things by his intelligent will-power. {D/K 21B, 23–25.}

This "one god" is identical with the world, or rather with the energy immanent in it. In other words Xenophanes made explicit the *pantheism* tacitly pervading Milesian thought. (Again, Anaximander had said that the Boundless "steers all things.") The three major Milesian insights—Unity, Immanence, Reason—are thus all reaffirmed in his worldview.

PYTHAGORAS

The second exporter of Ionian thought to Dorian Greece was Pythagoras of Samos, like Thales son of a Phoenician father and Greek mother, who arrived in the region at about the same time as Xenophanes—not however as a refugee from the Persians but "in disgust" with the regime of the tyrant Polycrates.

No records have survived for the history of mathematics in most of the sixth century B.C., but there must have been a lot going on, since some time in its middle or latter part Pythagoras proved the Theorem about right triangles that ever since has borne his name. He is said to have sacrificed a "hecatomb" (nominally a hundred oxen) in celebration of his feat. And very justifiably; it was the greatest single achievement of Greek mathematics, the foundation of trigonometry, and the prime illustration of the power of *discovery* inherent in the method of proof. Unlike the three theorems of Thales, Pythagoras's is by no means obviously true; and as far as we know, previously to the proof not even the Egyptians knew that it was true of all right triangles, but only of the 3–4–5 triangle and perhaps some other particular cases.

But Pythagoras was a many-sided character whose interests extended far beyond advanced mathematics. He adopted much of the Milesian teaching about the world, in the version of Anaximenes, and spread this new science to Croton inside the Italian boot. There he attracted some three hundred followers whom he united into a Brotherhood—in a way, the first scientific Institute—that in time took over the government of the city, though not for very long; their opponents corralled them in their meeting-house and burned it down with them in it. But Pythagoras himself and his organization survived the catastrophe.

Despite the name "Brotherhood," Pythagoras's institute was the only Greek gender-neutral organization ever. Women were admitted on equal terms, and Theano, Pythagoras's wife, was an eminent writer on her own, though all her works have been lost.

There were two degrees of membership: the *mathematikoi*, "learners," whence the word "mathematics," and *akousmatoi*, "auditors," who listened to Pythagoras's lecturing from behind a curtain but were not made familiar with the more esoteric doctrines. The organization as a whole was much given to secrecy, members being enjoined to observe *echemythia*, a Rule of Silence: if you are not certain what you should say, say nothing. Further, members who made discoveries did not claim individual credit for them, but attributed them to the Master, saying *autos ephē*, "Himself said . . . " For this reason it is impossible to be sure what Himself discovered all by himself, including "his" Theorem. (Some scholars doubt whether there was an individual "Pythagoras" at all. It matters little for us; after all, the discoveries did take place.)

According to a plausible legend, one Hippasos made the important discovery that the square root of 2 cannot be represented by any fraction both numerator and denominator of which are integers. Such a number—and there is an infinity of them—was, and still is, called "irrational"; their existence was a scandal to the Pythagoreans, and Hippasos, we are told, was murdered for disclosing it. The sphericity of the earth was known to the Pythagoreans, and some may have held that the earth itself rotates.

Besides mathematics and astronomy, Pythagoras pursued basic research in acoustics, establishing the correlations between harmonies, which are heard, and numerical ratios of the lengths of the strings that produce them, which are measured. This surprising discovery—numbers at the bottom of things—seems to have been what led him to declare that Things are Numbers. Since the Greeks sometimes represented numbers by arrangements of pebbles (like spots on dice), this theory may have been a forerunner of Atomism. Another mysterious pronouncement, that the world began "when the Unit inhaled the Void," seems to have been an attempt to combine this numerism with the Air theory of Anaximenes. (The Greeks thought of the Unit as the generator of numbers, but not itself a number.)

Alas, however scintillating the details, this kind of Grand Theory shows a falling away from two of the fundamental intuitions of Thales: first, that of the basic unity of things, the All is One. Manifestly the Pythagorean world is an irreducible plurality—indeed, several pluralities at once. Moreover, it comes from Nothing (the Void). This amounts to saying that there are unexplainable—"brute"—facts; a limitation is put on Reason. Heraclitus, back in Ephesus in Ionia, perhaps was unfair, but not by much, when he spoke of Pythagoras as "investigating many things" out of which he made "a wisdom of his own: *polymathie, kakotechnie*"—"much learning, bad art."

Even worse, from the Milesian standpoint, another important facet of Pythagoras's personality was downright mysticism and superstition— transmigration of souls, avoidance of beans, and the like. It was he to whom Xenophanes was referring in his story of the yelping dog. Where the idea of metempsychosis came from is unknown—certainly not Ionia nor even Egypt. Possibly India, where the similar ideas of *karma* and *nirvana* flourished, but more likely some intermediate region, such as Scythia. Anyway, it was Pythagoras's central conviction, and, paradoxically, motivated his science. All creatures that had soul in them, he held, are akin; therefore flesh-eating was cannibalism, and forbidden to his followers. Likewise beans. The aim of life was to *purify* the soul, which would then ascend through a hierarchy of reincarnations, ultimately to escape from the wheel of rebirth. What was touchingly Ionian about Pythagoras's outlook was his belief that the way to purify the soul was to do science—to get clear about the nature of things.

Pythagoras was possibly, after Thales himself, the most influential scientist and philosopher of all time, for Socrates and Plato, who well may have been card-carrying members of the Brotherhood, were adherents to his outlook: both sides of it, though, alas, more to the mysticism.

HERACLITUS

Probably the near-holocaust of Pythagoreans by the Crotoniates was not motivated by dissatisfaction with Himself's Grand Theory of Everything, but occurred merely because the common people "hankered after beans and revolted," as Russell put it. There were, however, two important thinkers who took issue with him on theoretical grounds: Heraclitus of Ephesus and Parmenides of Elea.

We have mentioned already Heraclitus's contempt for Pythagorean data-squirreling. This was not just a matter of taste, much less secret envy. Rather it was that Heraclitus, like Xenophanes, had an appreciation of the three principles of the new thought that Thales had initiated: Unity, Immanence, and Reason. When this is kept in mind, the alleged obscurity of the fifteen hundred words that have survived from his writings mostly vanishes.

"This cosmos," he proclaimed, "the same for all, no one of gods or men has made; but it was, is, and always shall be an ever-living fire, kindling in measures and going out in measures." {D/K 22B30.} What is out there is a *cosmos*, an arrangement, of *energy*, to use the modern word that best expresses what Heraclitus had in mind when he declared *fire* to be the basic stuff: the immanent source of *regular*—"in measures"—activity. The other "elements," earth, air, water, are "fire's turnings"—other *phases*, we would say, of this energy.

Pythagoras had discerned ten basic pairs of Opposites: left/right, unlimited/limited, even/odd, many/one, female/male, moving/resting, crooked/straight, dark/light, bad/good, oblong/square. The first members of the pairs are bad, the second are good. Right action, he claimed, would maximize the goods, minimize the bads, which he conceived as separable. Heraclitus to the contrary said that opposites are "the same," standing to each other in the tension that is another expression of "fire." He famously illustrated the point with the bow and the lyre: the frames want to go one way, the strings the other way, and this is just what makes them effective unities: in a way they "disagree with themselves," in another way they "agree with themselves." They can also be said to be "at war," but that is a good thing; if all war were to cease, an outcome for which the poets Homer and Archilochus had prayed, the cosmos would collapse.

Heraclitus's thought, however, is irreconcilable with Thales' on a major point. Since the opposites are "the same," they can never exist one without the other. This requires a "steady state" universe; it could not be the case that the diversity of things arose out of a previously homogeneous state. To say this is to deny the possibility of evolution and progress. Aristotle was the other great steady-state thinker of antiquity, and thus the principal reason why evolutionary thought, that got off to such a great start with Anaximander, eventually died out among the Greeks and was resuscitated only late in the eighteenth century A.D.

PARMENIDES

In Elea, on the southern west coast of Italy, lived Parmenides, a medical doctor by profession. He had been educated by a Pythagorean but was not a member of the Brotherhood. While still a young man he wrote an epic poem, no less, in which he expounded his own Grand Theory, taking off from criticism of the flaws (as he saw them) in Pythagoreanism, on grounds different from those of Heraclitus. Xenophanes probably visited Elea and may have influenced him, both in doctrine and in his choice of epic verse to expound it. The following is a conjectural reconstruction of his train of thought: it reproduces the conclusions that he reached, and *may* be somewhat like the reasoning that led up to them.

Some beliefs in the Pythagorean repertoire, he noted, constituted Knowledge with a capital K—they were truths that could not be doubted, such as the famous Theorem. Let us call these (Parmenides didn't) Class K beliefs. Other beliefs—Class O let us say—were open to doubt and revision in varying degrees. Now, he asked, what were these different classes of beliefs *about*? The objects of Class K beliefs were all of the same kind: quantities and shapes; whereas those of Class O were a miscellany of all the rest: the stars, developing embryos, gender differences, light, darkness, diseases, ... whatever. And the two classes were investigated differently: O, by eye, ear, nose, tongue, and fingers; K simply by thinking.

Now, Parmenides intuited, Class K truths—absolute, eternal, unchanging—must be *about* something itself eternal and unchanging. What could that be? Not the world that we see and hear and touch. Let us call it capital-R Reality. What can its further characteristics be? It must be all alike; Class K truths can't vary by place. It must be One and indivisible; Reality can't be chopped up into hunks and moved around—what could separate them? And of course ungenerated and indestructible. And since (so Parmenides assumed) only like can be Known by like, it must be a thinking thing.

What does all this add up to? Leaving aside the Thinking bit, *Space*. The Space of which the Pythagorean Theorem is true has all these properties, and nothing else does.

But if that is what Parmenides meant, why didn't he say so? Because it's not true that always "the Greeks had a word for it." They had words for place, position, location, void, and so on, but none for the absolute Space of geometry. He calls it *eon*, Being, or sometimes simply *esti*, IS, adding that "it is not possible for it not to be."

Now, what about the shoes, ships, and sealing-wax of Class O? Are they unreal? No; but they don't merit the capital R. They are changeable, knowable only a posteriori, not objects of scientific investigation in the strict sense. Nonetheless Parmenides himself took serious interest in them and made (or at any rate was first to state) important discoveries, such as that the Moon reflects its light from the Sun.

Parmenides cast his poem in the form of instruction given him by an unnamed goddess. It is in two parts, the first presenting "the unshaking heart of well-rounded Truth," the second "the opinions of mortals, in which there is no true confidence." (Note: *not* "no truth.") This is the distinction, here for the first time made explicit, between necessary and contingent truths. The former are defined as "what IS and is not possible not to be." The goddess prefaces her demonstration of the attributes of IS with the command "Don't use your unseeing eyes or echoing hearing or tongue, but judge by Reason (*logos*) my very controversial argument." "Judging by Reason" is what comes to be called the a priori method; eyes-ears-tongue, a posteriori. The a priori method, she contends, yields necessary truth, the a posteriori being capable of establishing only contingencies.

In arguing that IS could not be generated, for there could be no reason why, if so, this should happen at one time rather than another, the goddess appeals to the Principle of Sufficient Reason, or at least to Nothing from Nothing, which is its negative form. She makes no mention of propositions, nor does her discourse depend in any way on facts merely about language, so she (like all the ancients) is silent on the analytic/synthetic distinction (if there is such a thing). But with that exception, the goddess presents, explicitly or by clear implication, virtually the whole panoply of epistemological apparatus as it remains to this day. If philosophy is, as Renford Bambrough put it, "talk about what it is to be reasonable," then Parmenides was the first Philosopher, even though it was Pythagoras who invented the term. Parmenides' predecessors were the earliest Scientists (to employ a word coined only in the nineteenth century A.D.). Anyway, he certainly was the first Epistemologist.

Yet he was a scientist too, like the Milesians having his Grand—but not quite Unified—Theory of Everything. He reformulated their question from "What is the eternal basic stuff?" to "What *has to be* the eternal basic stuff?," thus ruling out of consideration the quasi-empirical methods by which the Ionians chose between Water and the rest. What it *has* to be is, as we have seen, Space-Thought.

Orthodox Pythagoreans made fun of the philosophy that declared Reality to be one, motionless, and so on. Parmenides had a disciple, Zeno of Elea, who wrote a book showing that the hypothesis of Many led to even more ridiculous consequences: that is, Zeno invented or at least popularized the form of argument known as Reductio ad Absurdum. He showed how, for example, a Pluralist, such as somebody holding that a line is constituted out of many distinct points, must as a matter of logical consequence deny that anything can move. But that is absurd!—a self-contradiction or at least a manifestly false assertion. Therefore the assumption of Many is false, too, since a truth cannot logically imply a falsehood. (Many scholars, perhaps a majority, still hold to the interpretation of Parmenides and Zeno as having "denied motion," that is, as having dismissed all motion as an illusion of the senses. For proof that this is a 180° misinterpretation, see the note at the end of this section.)

Parmenides had good news and bad news. The good news was his valida-
tion of the Milesian insight, that Reality is One and self-explanatory. The
bad news was that only the a priori disciplines, Mathematics and Logic,
were sciences, capable of establishing necessary truths; Physics, the study
of the many things that rattle around inside Reality, could only be "mortal
guesswork."

In hindsight we can see that if only Parmenides had been able to conceive of particular things as Modes of Space (Extension), his IS, there would
have been a happy ending: Spinoza's philosophy twenty-one hundred years
in advance! And he came *that* close, saying that "in relation to IS have been
given all the names that mortals have laid down thinking they were true:
birthing and dying, existing and not existing, changing places and alteration
of bright color." {D/K 28B8, verses 38–41.}

ZENO

Aristotle has preserved in summarized form four arguments "about motion"
(the orthodox say "against," but that is not Aristotle's word) by Zeno in
defense of his master's teaching. Each of his descriptions ends with a
paradoxical statement purportedly proved by what has gone before: that the
runner of a race cannot reach the goal; that Achilles (the fleetest Greek) could
not catch up with a tortoise, if the tortoise had a head start; that "the flying
arrow is at rest;" and that "double the time is half the time." Thus these arguments show (say the orthodox) that motion, being contrary to reason, cannot
really occur. Further, since they are arguments "against motion," and explicitly
offered in defense of Parmenides, Parmenides himself "denied motion."

Let us consider the most famous of them, "the Achilles." Suppose Achilles
can run a hundred times as fast as the tortoise, and they begin their race
with the tortoise 100 meters ahead. Then when Achilles reaches the 100
meter mark, the tortoise will be one meter ahead of him. When Achilles
reaches that point, the tortoise will still be one centimeter in the lead. And
so on ad infinitum, Achilles getting closer and closer to even with the tortoise but never exactly so, since to get there he would have to pass through,
one after the other, an infinite number of points (for between any two
points on a line there is a distinct midpoint), and it is of course impossible
to complete a task that has an infinite number of steps.

Mathematicians since Aristotle have tried to "solve" this "paradox" and
show that motion is logically OK, by (for instance) pointing out that if a
line is a dense series of points, so time is a dense series of instants, so that
the two series can be put into one-to-one correspondence. But we need not
consider whether they have been successful. For the earliest and best,

indeed almost the only, information we have about Zeno is given us by Plato in his dialogue *Parmenides*. There Zeno, twenty-five years the younger, is presented as having written a book the purpose of which was to "give back with something left over" to those who had ridiculed his teacher's *monism*. Plato lays great emphasis on this point, that all of Zeno's arguments were against *pluralism*, not "against motion." Parmenides' opponents then would almost certainly have been Pythagoreans, who are known to have held that "things are numbers," certainly a pluralism. If with this in mind we look again at the Achilles we can see what was going on. This argument does not start from nothing, but from the assumption that a line is an aggregate of points—an infinity of them. Zeno shows by reasoning *from this assumption* that motion cannot really occur. The argument is a reductio ad absurdum, the absurd consequence being that motion is impossible! That would indeed be paying the ridiculers of Parmenides back with something left over.

But if that was what Zeno was up to, one might ask, why didn't he say so? Well, that wasn't his style. Nor is it ours, usually. When we have shown that if your theory is true, then Einstein was an idiot, we don't bother to add, pedantically and anticlimactically, "But Einstein wasn't an idiot, so your theory is false."

A NOTE ON ELEATIC MOTIONS

(May be skipped by readers who do not believe that Parmenides "denied motion.")

Parmenides and his disciple Zeno are crucial figures in Greek science and philosophy. Unfortunately, the interpretation of their teaching is still a matter of controversy. They have been said to have "denied motion," not in some Pickwickian sense, but as meaning that, for example, rabbits don't really run, our opinion that they do being contrary to logic and therefore a delusion. Nor, really, are there any such things as generation, change, differentiation, and destruction. Real reality is just one homogeneous unmoving and unmoved thing, whether material or ideal being an unresolved question. This reading of the Eleatics is still widespread enough to be referred to as "orthodox."

This interpretation is impossible, both historically and philologically.

Historically: Two and a half millennia of playing with paradoxes: that all opinions are true, that nothing is known, that matter does not exist, that we make up space and time, and truth too, that we can't see trees, that nobody deserves anything, and so forth, have so raised the threshold of shock among "professional" philosophers that discussants of the Eleatics seem not in the least astonished by someone's "proving" that "rabbits don't run," not noticing that Nothing Moves (which amounts to Nothing Happens) and its immediate and obvious corollaries, There Are No Arguments, There Are No Arguers, You Don't Exist, *I* Don't Exist, is more paradoxical by several orders of magnitude than any on the short list above: indeed, a philosophical black hole, sucking in not only all of reality but appearance along with it. But attitudes during the first century of philosophy's existence *must* have been different: "Achilles can't catch the tortoise" *couldn't* have been anybody's actual or even "Pickwickian" belief; it could only have been the *absurdum* that something was being reduced to.

Nor were Parmenides and Zeno regarded by their contemporaries as crazy men. On the contrary, Plato spoke of "our father Parmenides" and apologized elaborately for having to take issue with him (not about motion). And Aristotle took him seriously enough to make one of his few excursions into mathematical topics to discuss the arguments about (*not* "against") motion, wishing to show that motion could take place *even if* a line was a "many," the premise that Zeno was reducing. And Simplicius, the sixth-century A.D. scholar to whom we owe the preservation of most of Parmenides' poem, remarked that "obviously Parmenides knew that he himself was generated, just as he knew that he had two feet when he said that the All is One."

Philologically: Let us scrutinize the texts involved, beginning with the poem of Parmenides.

It is in three parts: the first a Proem (preserved complete) describing the journey of the young poet in a chariot drawn by "very intelligent mares" to gates which maidens open to reveal an unnamed goddess (this much is agreed to be a conventional literary device) who greets him amiably and announces that she will instruct him about two things: "the unshaking heart of well-rounded truth" and "the opinions of mortals, in which there is no true confidence." The second part of the poem, preserved nearly complete and called (by editors) *On Truth*, expounds on what is the only way of valid inquiry, namely, that "it is and it is not possible for it not to be," and what conclusions may be drawn therefrom. After this the goddess begins the third section, *On Opinion*, saying "At this point I cease my trustworthy discourse concerning truth. Hear now the opinions of mortals, hearkening to the tricky ordering of my words." She is going to tell him this so that no (other) mortal can surpass him. Only a few, mostly short, fragments remain from this part. They take up a miscellany of topics from gender differentiation in the embryo to the Moon as reflecting light from the Sun.

Now to interpretations. Orthodoxy takes the contrast between On Truth and On Opinion to be starkly that between true and false, since the goddess says that there is no "true confidence" in the latter. But this expression is not just poetical for "false." The Opinions are patently of the sort that are not *eternal* truths, nor guaranteed against error and/or revision, but that only means they are (as we have put it in our text) a posteriori, warranted by fallible—but not necessarily erroneous—experience. A cut below mathematical truth, but not to zero.

Astonishingly, the case for the ascription of the unreality of motion and change rests entirely on how a single ambiguous Greek word of three letters is to be read in its context. The word is τωι—tau, omega, iota—which is basically the dative case singular of the relative pronoun, "for which"; but it also can mean "therefore." Orthodox translations give it the latter signification in a certain crucial passage, Fragment 8, lines 36–41, rendered by Tarán as

> nothing other, besides Being, either is or will be, since Destiny fettered it to be whole and immovable; therefore, all that mortals posited convinced that it is true will be [mere] name, coming into being and perishing, to be and not to be, change of place and exchange of brilliant color. {Tarán 1971} ("[mere]" is Tarán's addition; there is no equivalent in Parmenides' text.)

The sense of the passage then, the orthodox say, is that words (including "be"!) presupposing real change are nothing but meaningless noises. If, however, τωι is read as "for it," the added "[mere]" is expunged, and "it is" corrected to "they are," we shall have

> ... nothing other, besides Being, either is or will be, since Destiny fettered it to be whole and immovable; for it all that mortals posited convinced that they are true will be names, coming into being and perishing, to be and not to be, change of place and exchange of brilliant color.

The sense will then be that words presupposing change nevertheless ultimately refer to that which does not change: Reality as a whole. Rabbits, then, spatial creatures that they are, really run, but what they run *in*, space, does not move. This, I think, makes better sense. So far from being wildly paradoxical it turns out to be a truism—which was not, to the Greek way of looking at things, a disadvantage.

{For more extended treatments of these topics see Matson 1980 and 2001, both reprinted in Matson 2006, and bibliographies therein.}

EMPEDOCLES

Empedocles of Acragas (modern Agrigento) on the southeast coast of Sicily, the second of the native Dorian philosophers and a Pythagorean, defended Himself against Parmenides' critique. He tried to vindicate the scientific status of physics, the study of particular moving and changing things, by arguing that in a way they are eternal: they are mixtures of four "roots," earth, air, fire, and water, the great obvious stuffs of the world, later called elements, which *are* eternal. Each root has, or rather, *is,* essentially four of the sixteen possible qualities obtained from the pairs bright/dark, hot/cold, thick/thin, and wet/dry; fire for example is the Bright-Hot-Thin-Dry. Mixtures of the roots, he averred, in various proportions could account for all the various qualities of things we perceive, just as all the colors can be obtained by mixing only a few pigments. He even gave recipes: bone, for example, = 1 part earth + 1 part water + 2 parts fire.

What mixes the roots is an external force, Affection; what separates them is Strife. They operate cyclically. When Affection is supreme, all the roots are mixed into one homogeneous sphere; when Strife rules, all are separated into concentric layers. Two kinds of animal evolution go on: when Affection is gaining over Strife, body parts are spontaneously generated:

> Many heads grew up neckless, and naked arms wandered around lacking shoulders, and eyes rambled destitute of foreheads.

Affection unites these fragments at random; only those few survive that are capable of coping with the world. During the opposite cycle, when Strife is gaining, "undifferentiated creatures" arise, "not yet displaying the lovely form of limbs, nor the voice and organ appropriate to men." Strife works further to split these bisexual spheres into men and women, hetero- and homosexual. Human love is the longing and search for one's literal other half.

This magnificently imaginative account is, however, tethered to very few low beliefs. It contains the idea of survival of the fittest, but in a form inferior to Anaximander's. And none of the three Thalesian insights—Unity, Immanence, *Logos*—is to be discerned in it.

As for the soul, Empedocles is an orthodox Pythagorean, a believer in transmigration:

> For already once I was a boy and a girl and a bush and a bird and a dumb salt-water fish.

But there is a problem: How can the same soul survive to migrate, if every particular thing is an unstable mixture of the four roots? The only plausible answer, not found explicitly in the fragmentary texts, seems to be that the soul is a bit of pure Affection, of which this poet-doctor-prophet-prince sang:

Look upon her with your mind, do not sit with eyes in a daze: her we recognize as implanted in the parts of mortals. By her influence we think kind thoughts and work in concord, calling her by the names of Joy and Aphrodite.

SUMMARY

Xenophanes of Colophon, who had "heard" Anaximander, brought Ionian science to Magna Graecia (Sicily and southern Italy) when he fled from the Persian invaders of his homeland in the middle of the sixth century B.C. He made geological discoveries implying the cyclical nature of cosmic time. He denounced the Olympian religion as both silly and immoral. But he had his own conception of "one god," who does not move, but moves everything else by "intelligent will-power." This Being is really the All of Thales, the immanent energy (as we would say) of the cosmos: Pantheism.

Pythagoras, an immigrant from Samos to the same region at about the same time, carried mathematics to its highest point in antiquity with his famous Theorem. In Croton in the Italian boot he set up his Brotherhood—a sort of combination of research institute and religious order. He investigated natural science, notably in the application of mathematics to music, leading to the pronouncement that Things are Numbers. However, he was lax in observance of Thales' requirements of unity and reason in his Grand Theory, and even introduced mystical and superstitious elements, such as the transmigration of souls. However, it was this central belief that motivated his science. For it made the aim of life to be to *purify* the soul; and Ionian that he was, he thought the best way to do this was to get clear about the nature of things.

Heraclitus of Ephesus, back in Ionia, criticized Pythagoras with respect to unity and *logos*, reemphasizing their importance for the new way of thought. His choice of Fire as the basic stuff amounted to making immanent and unifying Energy the central scientific concept.

Parmenides of Elea also took his starting point from criticism of Pythagoras. He was first to make explicit the basic pair of distinctions in the theory of knowledge: necessity/contingency and a priori/a posteriori. He held that what necessary and a priori knowledge—mathematics—is *about* is Space, the eternal and unchanging reality; the things *within* that reality can be studied only by empirical methods: "mortal guesswork." There can be no science, strictly speaking, of Nature. Hence his Grand Theory was a dualism, rejecting the Thalesian requirement of unity.

Empedocles of Acragas tried to vindicate Pythagoras's science by declaring that individual changing things are mixtures of four Roots, in themselves eternal and unchanging, hence fit objects of a priori knowledge. He held that the world goes through endless cycles in which now Affection, now Strife (external forces) rule. He posited two differing kinds of animal evolution as occurring in the intervals when one force is gaining on the other. Nevertheless, as an orthodox Pythagorean he believed in the transmigration of souls, which therefore must not be mixtures of the Roots, but perhaps bits of Affection. His imaginative system, tethered to few low beliefs, rejected all three of Thales' requirements, being neither monistic nor naturalistic nor rationalistic.

CHAPTER 11

✦

Athens I

L iteracy, in an alphabetic writing adapted from their neighbors the Phoenicians, came to the Ionian Greeks only about a century before the time of Thales. Immediately the Homeric poems *Iliad* and *Odyssey*, products of a long tradition of oral composition, were written down: the first works of European literature. Alongside them flourished a new literary genre, the lyric poem. Short, intensely personal and emotional, only meager fragments of it from Anacreon, Archilochus, Sappho, and others remain to tantalize us. And although hymns to various divinities were composed, on the whole this was a secular literature. This is true even of Homer: the gods of the *Iliad* are only people writ large, often behaving so deviously and from such petty motivations as to provide the poem's comic relief. In the *Odyssey*, perhaps redacted by another hand, they exhibit some moral improvement but remain far from being pillars of righteousness. There is an afterlife, but dull and miserable, no reward for virtue; the shade of Achilles declares that he would "rather be the hired man of a landless share-cropper on earth than King in the realm of Hades." In Ionia, so untypical of the ancient world in lacking a caste of professional priests, the connection of religion to morals was both tenuous and ambiguous. This is a main reason why the new scientific way of looking at things, which, as Xenophanes pointed out, made a laughingstock of popular religion, was propagated for nearly a century before the first attempts to suppress it were made. Others were the importance of persuasion (therefore of argument, therefore of logic) in cities without absolute rulers; and Greek curiosity, typified by Herodotus, and adeptness at collecting and classifying facts, as by the medical writers.

We have seen, however, that high beliefs play an important role in all human existence and that challenges to them are, for good evolutionary reasons, always resisted. So it was inevitable that the developments in Ionia would produce an intellectual reaction. It occurred first in Athens, dominant city of Greece after the repulse of the Persian invasions.

Ionian science spread to Athens later than it did to Italy, even though the Athenians were Ionian and the Italians were Dorian. Aspasia, a Milesian lady, became mistress of the Athenian leader Pericles. She patronized Anaxagoras, an immigrant from Clazomenae in Ionia, who was in the direct line from Thales, although his Grand Theory departed as far as was possible from the requirement of Unity. He held that there were as many basic stuffs, "seeds," as there were discriminable qualities. All apparent change was mixture and separation of them. "How can flesh come from what is not flesh," he asked, "or hair from what is not hair?" To the limited capacities of our senses things appear to be "pure" this or that according to which seeds predominate in the mixtures. (But it is even more complicated than that. Even a hair seed, could it be isolated, would not be "pure" hair, but a fusion in which the hair quality predominates.) The agent of mixture and separation, however, is Mind, the only unmixed thing, which operates to create order: the *logos* part, which again, contrary to the Thalesian requirement of immanence, becomes an external force. Apart from this unsatisfactory theory, however, Anaxagoras was a thinker of great brilliance, the first to speculate that there are worlds like ours elsewhere in space.

His pronouncements that the Sun is "a red-hot stone, bigger than the whole Peloponnese" and the Moon is earth—not gods—got him into trouble. Conservative enemies of Pericles, out to get their political foe by indirect means, denounced him as a blasphemer. He was tried and condemned to death. However, he escaped to Lampsachus back in Ionia, where he finished out his life as a revered schoolmaster.

It was a quarter of a century later that this opening salvo in the warfare of science with religion was followed by another episode. In the meantime, Athens witnessed the establishment and flourishing of a new profession or industry, that of the *sophists*.

The curiosity so characteristic of the Greek people, after the tremendous impulse given to new ways of thinking initiated by Thales and his successors, soon led to investigations into diverse fields far removed from mathematics and "physics." And a way was found to make money out of the new interests. The Sophists ("*Sophistes*" = "wise (man)," skilled in some art or subject; "wizard" would be its exact English equivalent, had that word not acquired its connotation of involvement in magic) were or purported to be experts on everything connected with *logos*—not only science but grammar, rhetoric, and the arts of persuasion. They typically went from city to city, delivering public lectures and providing private tutelage for fees, often high ones. The stock complaint against them was that they taught their clients how to "make the weaker cause appear the stronger," without regard for truth. And indeed the familiarity with divergent mores that they acquired in their peregrinations did sometimes lead to oversophistication.

The most famous sophist, Protagoras, developed for the first time an explicit Relativism. Each person (he held) necessarily perceives the world from his particular point of view; there is no way to get outside these perspectives and judge one to be more correct than another. "Man"—the individual man, not the species as a whole—"is the measure of all things: of things that are, that they are; of things that are not, that they are not." In other words, "what seems to a man is to a man." Even worse, he wrote a book *On the Gods* beginning "As to gods, I don't know whether there are any, or what they might look like. For many are the obstacles to knowledge: the obscurity of the matter, and the shortness of human life." (It must have been a very brief book.)

And then there was Socrates. As a young man he had an interest in the new thought, and even took lessons from the sophist Prodicus, whose specialty was the right use of words. But he seems to have had some sort of conversion experience after which he conceived of his mission as the moral one of persuading the Athenians to revaluate their values, to get straight about what made a life worth living—the *logos* of ethics. In pursuit of this aim he was in the habit of publicly questioning people's conceptions of good and evil, a process often embarrassing to his interlocutors though amusing to the young men who hung around him. But he claimed to have no wisdom of his own, only to be seeking to elicit it from others. Rightly or wrongly, he was suspected of being an associate of, or at least in sympathy with, the Pythagoreans, who had so disastrously pursued a policy of political reform in Croton.

Upon this scene came the greatest comic poet of all time, Aristophanes, who in 423 B.C. produced his play *The Clouds*. It takes place in the *Phrontisterion*, "Thinkateria," the teaching and research institute run by Socrates, who makes his entrance in a basket high in the air (Anaximenes!) and ridicules belief in the existence of Zeus, declaring that the Whirlpool is now "king." The inmates of the institute, skinny pasty-faced nerds, are investigating astronomy, (Anaxagoras!), measuring flea jumps (Pythagoras!), geology (Anaximander!), grammar (reform of ordinary Greek to make it more logical!—the sophist Prodicus), meteorology (thunder as analogous to stomach-rumbling after overeating), and especially rhetoric. The old man who applies for admission to the Institute in order to learn how to cheat his creditors is found to be ineducable, so he sends his son, who after being instructed in argumentation by the Unjust *Logos* himself, beats up on his father (very no-no behavior to a Greek), then cleverly "proves" that he was right to do so, and threatens next to deal similarly with his mother. The old man, now sobered, gathers his cronies and they set the Thinkateria afire with Socrates and his crew still in it (the conflagration of Croton!). That is what they deserve, they shout, for denying the gods.

We are told that Aristophanes and Socrates were on friendly terms personally, though at his trial twenty-four years later Socrates partially blamed popular hostility to himself on the play, which, he averred, treated of topics with which he had no concern. No doubt he was right about the situation at that time; he had moved on to Ethics, as we have pointed out. But the play, which was largely the cause of Socrates' celebrity in Athens and throughout Greece, was what most people "remembered" about him. The indictment against him read that he "does not honor the gods that the city honors, but introduces new divinities"—the Clouds!—"and he corrupts the youth," exactly what the play depicted him as doing.

Though Aristophanes rightly considered *The Clouds* his best play, it was a flop—too intellectual, and not obscene enough, for a mass audience. It showed beyond doubt, however, that the new thinking from Ionia with its rejection of high beliefs had provoked opposition, and not just from ignoramuses but from people with the intelligence of Aristophanes. But this reaction had nothing to put in its place and came too late for its forcible suppression. As things turned out, the really heavy—and largely successful—attack on it came from a *disciple* of the later-period Socrates: Plato.

SUMMARY

The intensely human culture of newly literate Ionia spread into every field of intellectual endeavor. Though it was obviously anti-religious, serious opposition to it took nearly a century to develop.

Ionian thought came to Athens later than to Italy. It was introduced by Anaxagoras of Clazomenae and by members of a new "profession," the Sophists—itinerant lecturers and tutors. The most famous of these was Protagoras, the first Relativist and explicit agnostic.

Socrates, a native Athenian, started out as a friend of the scientific side of Pythagoreanism. As such he was caricatured by the comic poet Aristophanes; and as such he was condemned and put to death for "impiety." But by that time he had undergone a conversion from science to the moral and religious interests also associated with the Brotherhood.

CHAPTER 12

∘∿∘

Atomism

As everyone knows, less than two centuries after the beginning of science certain Greeks came up with the atomic theory of matter, a basic constituent of the true theory of the universe, virtually lost and revived (by John Dalton) only near the end of the eighteenth century A.D. This astounding achievement is often treated as just a lucky guess. {See, e.g., Russell 1945} Unless all the main contentions of this book are mistaken, that was not at all the case. It took genius to formulate the theory; and yet it can be seen in hindsight as the inevitable next step, after Anaxagoras, in the search for the unified, naturalist, and rationalist basis of things.

The originator of the theory was one Leucippus—of Miletus!—who formulated the basic Thalesian insight in its most elegant version: "Nothing happens at random, but everything for a reason and by necessity" (*ek logou te kai hyp' anankes*). But nothing further is known of Leucippus, not even his approximate dates, except insofar as he must have been somewhat older than his pupil Democritus, who was born about 460 B.C., in Abdera, a northern Greek city founded by Ionian refugees, hometown also of the sophist Protagoras. Democritus was wealthy and traveled very extensively. He founded no institute of his own, which no doubt is the main reason why although he was perhaps the most prolific writer of antiquity—catalogues of his works list more titles even than those of Aristotle—not a single connected treatise by him has survived. About three hundred fragments remain, but all are brief and all but a few are about morals, not the physics and mathematics in which he was preeminent. However, Atomism is so simple and elegant a theory that this regrettable fact gives rise to few serious difficulties of interpretation. Furthermore, it was presented again, more or less intact, by Epicurus in the third century B.C. and by the Roman poet Lucretius in the first.

We have seen how speculation about the basic stuff of reality came to a dead end in Anaxagoras, who postulated as many different "seeds" as there are differences to be accounted for—thus explaining nothing. Democritus perceived what had led to this impasse: the assumption made by all previous investigators that matter could only be described *qualitatively*, as hot or cold or wet or dry (the favored qualities), or as in Anaxagoras, these plus the hairy or the fleshy or . . . and so ad infinitum. Democritus rejected this assumption.

The most important of the Democritean fragments goes: "By custom color, by custom sweet, by custom bitter; in reality atoms and the void." The distinction between what is natural to all human beings and what is the result of human conventions based on agreement was a major topic of debate in the fifth century B.C., mainly in a political context: for example, the sophist Antiphon held that all human beings are equal; slavery and other hierarchies are only "by convention." Democritus is here extending this dichotomy to fundamental metaphysics, but with a difference. He does not mean that whether something is sweet or sour is a matter that has been decided by a treaty, as it were, but that nothing is either sweet or sour in itself; tastes are *interactions* between a thing and a perceiver: "in reality, atoms and void." The atom, "uncuttable," that is, smallest particle of "the full," has, and has had for all eternity, all by itself *only* the properties of size, shape, and weight, and is besides in motion, the direction of movement being a consequence of the most recent collision that it has undergone. It has by itself no color, odor, or taste. In a modern vocabulary, real things have *objectively* only those properties that are conserved through change: shape and weight. Colors, smells, tastes are *subjective*; they are not conserved, and they vary with the perceiver.

We can see in this teaching a reply to the relativism of Democritus's fellow-townsman Protagoras: Yes, if it seems hot to A and cold to B, it *is* both hot and cold; but now that these terms are properly understood, they no longer pose a threat to scientific objectivity.

Basically, all that atoms can do is move, and hence collide with other atoms. But atoms have all sorts of shapes, including hooks that can join up with other appendages on other atoms to form large-scale, visible objects. And the motions of atoms not being random but in accord with necessity, all these collisions, and the objects resulting from them, are in principle calculable, that is, they do not result from chance. But neither—contrary to Anaxagoras and (later) Aristotle—do they move *in order to* bring about anything in particular. Thus Democritus absolutely eliminates teleology from his system. Everything is necessarily determinate cause producing necessarily determinate effect. And everything can in principle be explained this way: "mechanically," we say, for this is the way we already know—

having built them—that machines operate. This view, that given a cause the effect follows necessarily, is called Determinism.

An ancient story, whether true or not, illustrates the difference between this mode of explanation and god or mind-based accounts. The tragic poet Aeschylus reportedly died when, as he was strolling on a beach, an eagle flew overhead and dropped a tortoise on his head. A curious fate, calling for explanation. Traditionalists had theirs when it was found that an oracle had predicted that he would die of "a bolt from Zeus." Though this had been thought to refer to a lightning bolt, the facts could also fit this strange event, since the eagle was sacred to Zeus. Democritus, however, went to the beach and made some observations. He noted that eagles were fond of tortoise meat, but in order to get at it, they had to crack the carapace. They did so by grasping the tortoise in their talons and dropping it from a height onto a rock. This habit, together with the fact that Aeschylus was bald, explained to the materialist philosopher how and why what happened had happened.

Aristotle complained that this mode of explanation amounted merely to saying in every case, "Thus it happened formerly also," which so far from explaining anything, multiplied the need for explanations. But this criticism misses the mark. The twofold purpose of explanation is to remove puzzlement (which Democritus's did) and to fit an apparently anomalous occurrence into the underlying regularity of nature, thus showing, or tending to show, that it "could not have been otherwise," an aim that the Philosopher endorsed.

Democritus said that he would "rather discover one causal explanation than be Shah of Iran," but (apart from the death-of-Aeschylus case, if it really happened) he was notably unsuccessful. He even retained belief in the old Ionian picture of a disc-shaped earth, long outdated by the Pythagorean discovery of its sphericity. But that was not his fault. The atomic theory, at this time, was tethered at a very high level, and much more—in particular, laboratory equipment never available in the ancient world—was needed if the ropes were to be effectively pulled in.

Democritus was aware of some of the methodological problems he faced. Being a scientist, he realized that a sound theory had to be either implied by or tethered to low beliefs, ultimately therefore by sense perception. But how could this be? If nothing really happens but atomic collisions, then perception itself must be atoms colliding. When one perceives an apple, in some way atoms from the apple have to be initiating a series of collisions along a path through the eyes, into the brain, ending up hitting the soul atoms. How could that give information about an apple? Further, perception seemed to reveal only what according to his primary insight was "by custom"—colors, odors, and so on. These reflections sometimes led him almost to despair: "In reality we know nothing. For truth is at the bottom of a well." But he pulled himself back up by the insistence that Reason is a different and finer "instrument for knowing." We do not know how he defended this contention.

EPICURUS

Atomism might have vanished altogether, like Democritus's writings, had it not been taken up near the end of the fourth century B.C. by Epicurus, a native Athenian whom events had caused to grow up in Ionia, where he had to assist his mother in making her living by casting spells and such-like chicaneries. These unpleasant experiences instilled in him a lifelong disgust with superstition and religion, between which he made no distinction. (There isn't any.) From a certain Nausiphanes he learned about Atomism, in which he found just what he was looking for, a worldview with no spooks. He claimed, however, to have worked it out for himself.

As far as we know, Democritus had no ethical axe to grind but developed his philosophy from a disinterested desire to find the truth. The difference between his and Epicurus's motivations explains the two modifications that the latter made in Democritus's system: (i) Whereas Democritus had denied any beginning or preferred direction of motion to the atoms, Epicurus held that the natural motion of all atoms is downward, at the same speed, and (ii) that, rarely, an atom may "swerve," just a little bit. "Down" Epicurus defined as in the head-to-foot direction. Now in an infinite universe there can be no preferred direction, nor, if everything moves at the same speed, is there any difference between their motion and their rest. But Epicurus did not notice these points. So, in effect, the Swerve was the first motion (an atom changing its position relative to another atom).

The reason Epicurus gave for making these modifications was that if the motion of every atom is strictly determined, then everything happens of necessity, or as he put it, by Fate. But, he held, human beings are not thus constrained; they can make choices, they have what later came to be called Free Will. Otherwise it would be pointless to try to save them from religion or anything else.

This seems to be the first appearance in philosophy of the Free Will Problem, on which more ink has been spilled than on any other of the traditional philosophical puzzles.

Epicurus returned to Athens, bought a plot of land, and set up a sort of philosophical commune called the Garden. It persisted into later antiquity, being one of the three official schools of philosophy, along with the Porch (of the Stoics) and the Academy (of the Platonists), later subsidized by Roman emperors. Julius Caesar was said to have been a closet Epicurean; the poet Horace was more open in his allegiance.

The Roman poet Lucretius in the first century B.C. wrote *De Rerum Natura*, "On the Nature of Things," an epic poem in six books expounding the philosophy. It got through the Middle Ages to modern times in a single manuscript.

SUMMARY

Atomism, the high point of ancient science, was worked out less than two hundred years after the beginning of science. It was not a lucky guess but a consequence of previous investigations.

Leucippus of Miletus was its originator. More is known of his pupil Democritus of Abdera who gave it its classical formulation, that Reality consists of atoms moving in the void, and of nothing else.

The Democritean breakthrough was the realization that the basic stuff need not—indeed, could not—have objectively, by itself, the qualities of color, heat, cold, wetness, dryness, taste, and smell, for these are not properties but *events* that happen when an object and a perceiver interact. All that the atoms have by themselves are the properties conserved through change, namely, shape, size, and weight; and they are naturally in motion from eternity. They are capable of joining together to make large, visible objects. Their motions are the necessarily determinate consequences of the collisions that they have undergone, which are in principle calculable. This is Determinism.

Democritus was troubled by methodological problems. His science's credentials required support from low beliefs, namely, perceptions. But by his analysis, perception is only a "flow toward us" of atoms—how this can get us in touch with the real things was obscure. He brought in Reason as a "finer instrument" to solve this problem—how it did, we do not know.

Epicurus adopted atomism as a science justifying rejection of religion and superstition. In line with this purpose he modified the Democritean principles to allow for a "swerve" of atoms, transforming the science from a strict determinism into one which (so he wrongly thought) by allowing some wiggle room for atoms made possible Free Will—a major philosophical problem that here becomes explicit for the first time.

Epicurus founded, in Athens, the Garden, a philosophical commune that became (with the Platonic Academy and the Stoic Porch) one of the official Athenian schools of philosophy.

Lucretius, a Roman poet of the first century B.C., wrote *On the Nature of Things,* an epic poem in six books, which transmitted a full account of this science and philosophy to modern times, in a single manuscript copy.

CHAPTER 13

cᐱↃ

Athens II: Plato

If our distinction between science and philosophy is accepted, Parmenides must be recognized as the first philosopher. But he wrote only one work, a poem of a few hundred lines, the strictly philosophical content of it being chiefly the setting out of the basic distinctions necessary/contingent and a priori/a posteriori. Plato in his lifetime of eighty years wrote nearly two thousand pages, which include *The Republic,* indisputably one of the two greatest *single* philosophical works. (The other: Spinoza's *Ethics.*)

Born of an aristocratic and influential family, Plato was expected to pursue a political career, and his interest in the reform of governance was heightened by his association with Socrates in the latter's later years. But like his guru, he was contemptuous of the very idea of democracy, the rule of the ignorant. Reform could come only from the top. But then, in the sorry events after Sparta imposed oligarchic rule on defeated Athens, two of his own relatives proved to be not the least horrid of the Thirty Tyrants, as they came to be called. Plato came to the famous conclusion that the only hope for betterment of the human condition lay in the bare possibility that philosophy and power might somehow be united in the Philosopher King. To educate young persons who had the innate qualities needed for such a post, he founded and devoted the rest of his life to the Academy, the first university, which continued in existence for nine centuries.

The curriculum heavily stressed mathematics: plane and solid geometry and theory of numbers. This might strike us as odd basic training for rulers, but Plato had his reason: the Philosopher is one who Knows. Agreeing with Parmenides that mathematics was the only genuine Knowledge so far attained, Plato sought, as the Eleatic thinker had done, to understand why this was so. And like Parmenides he found the answer to lie in the unchanging eternity of its object. But that object was, for Parmenides, the One Space. Plato, however, descended one level of abstraction, as it were: instead of space itself as the object, he perhaps more commonsensically held the spatial figures—triangles, circles, and so on, a Many—to be the basic subject matter. But, of course, not this or that triangle, sketched in the sand and erased at the end of the class; rather, The Triangle, an object seen not with the body's but with the Mind's eye; that of which the figure under foot was but a crude reminder, drawing what little degree of Reality it had from its Participation (a relation never explained) in the eternal and unchanging Idea.

So far so good. There is intuitive support for this notion of geometry's being about abstract objects understood rather than (physically) seen. But then what?

Whether or not he was conscious of the fact, Plato was the first thinker to have a Philosophy of Language at the bottom of his ontology, his theory of Being or Reality. He did not consciously ask himself the question, How do words *mean* what they mean? He took for granted that he already knew the answer. Words like Socrates—what we call proper names—are labels that we (metaphorically) paste onto things. *And so are all other words.* Words mean objects by being their names. If a word has a meaning, then there must be some real thing that it is the name of. So, "Triangle" is the name, not of the scratch in the sand box, but of the eternal, unchanging object that the Mind perceives when it does geometry. But this is a general fact about language, so Plato thought: man, cat, dog, fatherhood, good, justice, virtue, . . . you name it, are equally names of eternal and unchanging objects seen by the mind's eye: "Ideas," he called them, or "X in itself," or "the very X." Knowledge of them and of their properties is intuitive, not experiential. But how to intuit them, how to prepare for the "Aha!" experience that one has on "seeing" the cogency of a proof, is an ability that only a few have, and which can be developed only by a long and arduous process, beginning with mathematics (for ten years) and then five years of "dialectic," questioning by and arguing with those already in command of the technique. Even then, fifteen years more of "practical" experience in perfecting and relating this Knowledge to the actual affairs of the people must be undergone before the candidate is in a position to intuit the highest Idea, that of The Good; after which he (or she; Plato allowed the possibility of Philosopher Queens) will be ready to rule according to Real Justice and so on.

That is a brief sketch of Plato's scheme for training the people who would be ready to take the reins of government should it some time happen that an heir to a tyranny might prove to have a philosophical nature and find himself in power. The cure for the evils besetting his subjects would be drastic: it would require, for example, the banishment from his city of all above a certain age as incurable because of having grown up under the old corrupt regime; but in principle it could be brought about.

In his writings, which are almost all dialogues in which Socrates "adorned and rejuvenated" is the principal figure, Plato ascribed this theory of Ideas to that famous man. We can be sure, however, that in doing so he was paying homage rather than reporting. While it is true that Socrates was concerned with exact expression, both on his part and on that of his interlocutors, he had no theory of language. And it seems natural that the very first theory of what it is for a word to mean would be this simple one, that meaning is naming. After all, the grammar with which we are familiar, with its distinctions of seven "parts of speech," was still a century or more in the future. (It was an achievement of the Stoics.)

"Meaning is naming"—to Plato no doubt not a theory at all but simply an obvious fact—became the point on which the philosopher erected an inverted pyramid of *metaphysics*. (There seems to be no way to avoid anachronistic use of this word.) Like the world of "our father Parmenides," Plato's is a duality of intelligible (really Real) and "sensible" (the things seen through the body's eyes). But while Parmenides' Reality is the One Space, Plato's is a vast warehouse of Ideas, quite distinct from one another; one per meaningful word. The relations between them are mysteries that the author tried through his life to solve, with little success. However, they certainly form a hierarchy, with the Idea of the Good at the top. Do they do anything? No, they are eternal and unchanging; they just sit there, being "participated" in by the not really altogether real things of the sensible world. Why *is* there a sensible world at all? Plato's answer to this question is given in the dialogue *Timaeus*, in which "a stranger from Elea" explains that there exists (or existed) a superhuman being, simply called The Worker (*demiourgos*) who, impressed by the excellence of the Ideas, wanted to make as many copies of them as possible. But as he had nothing to work with but base matter, necessarily his copies were imperfect, changeable, and perishable. Whether Plato intended this story to be taken as the literal truth, or only as a "myth," is uncertain; more likely the latter.

The effect of this philosophy, if accepted, would be to reinstate high beliefs to the supreme status they had enjoyed before Thales. True, they were a new kind of high belief—entities of reason intuited and hypostatized by specially trained intellectuals—but high beliefs nonetheless. The rules of Plato's game precluded putting any Idea to a test in experience. If the "sensible world" did not comport with the Idea, so much the worse for the sensible world, which was mere appearance, not really real. Most significantly, high-belief categories of explanation—purpose, will, goodness—were reinstated.

Plato could carry out this program because he was able to exploit the general failure of Greek science to test its hypotheses. Although this failure was excusable—after all, direct testing of the atomic theory of matter, so brilliantly excogitated by Democritus, could not be done until almost our own times—the fact is that logical consistency was practically the only check on Greek hypothesizing. And Plato was careful, on the whole, about logic, besides mathematics the only low-belief element in his system.

All three of the requirements that Thales put at the basis of science are unsatisfied in Plato's Grand Theory. Unity is ignored altogether. This is a two-story universe with, on the second story, an infinity of postulated entities. The second requirement, that of immanence—changes explained by Natures, internal principles—likewise is not to be found in it. The third, *logos*—Sufficient Reason for things being as they are—is also entirely missing; the silly story of the Demiurge cannot satisfy it whether taken literally or figuratively. Indeed, the entire account represents a reversion to pre-Thalesian thought.

Plato, by forcing an unnatural union between logic and high beliefs, invented *theology*, a mimicry of scientific reasoning that superficially makes high beliefs look as if they have scientific support. He inaugurated this enterprise with his usual brilliance in the tenth book of his *Laws*. High believing, before the beginning of science, did not feel itself constrained by logical considerations, as has been pointed out; for it is the impossibility in practice of holding (and therefore of acting on) contradictory low beliefs that is at the basis of logic. We are not born, prior to experience, with the logical rules hardwired in; whatever their ultimate source, it is experience that validates them for us in the first instance. But the gods are exempt from logic. They can be eternal and be born; they can take this form and that, and at the same time; they can be here and also there ("omnipresent") simultaneously; and they can know everything without having learned anything. It is not, however, much of a sacrifice, from the religious standpoint, to incorporate logic. Logic imposes few limits on imagination, and anyway, in case it is inconvenient, it can be got around by various stratagems. And incorporating logic gave reasoning about high beliefs the appearance of intellectual rigor, which from the beginning was associated with logic. Logic is, after all, the backbone of science—of all science, but especially of mathematics. Theology sometimes adopts allegedly self-evident propositions as axioms, in direct imitation of geometry—for example, in *Laws* Book X, where a whole slew of axioms are announced, from which the existence of gods is "proved."

The pretensions of theology to be a science continue to this day to be persuasive, at least to the extent that many anti-theologians concede that it is reasonable to be challenged to refute the arguments. But it is pointless to try to reason, logically and scientifically, with the high believer. For theology is not a variant of science, any more than a rhinestone is a variant diamond; nor is it in any way the product of the same basic intellectual impulse.

Although religion and high belief are not identical—there are political, economic, and ethical high beliefs, and there are low beliefs that must count as religious, for example, that Jesus lived in Palestine during the first century A.D.—the central beliefs of all religions are high. Religion therefore has a different epistemology from science. And insofar as religion and science make pronouncements about the same things, for example, origin, age, and development of the cosmos, they do so on entirely different grounds, and, as far as the truth of the contentions are in consideration, it must be war to the knife between them. All attempts at compromise, such as assigning different spheres—body and soul, nature and grace—are misguided and bound to fail, for the simple reason that a high and a low belief about the same subject cannot coexist.

It is possible to look at Plato as a would-be reconciler of science and religion; after all, his theology, as just pointed out, did make considerable concessions to scientific thought. Aquinas, Descartes, Leibniz, and Berkeley are a few of the thinkers who have regarded the "reconciliation of science and religion" to be their task, and who have expended astounding ingenuity on it. It was, however, wasted effort, for the two strains of belief involved are in the strange relationship of being really incompatible, even though one of them, the scientific, owes its existence (historically and psychologically, not logically) to the development of imagination in human brains that manifested itself in the human propensity to high believing.

Plato was well aware of the importance of systematic organization in a worldview. The Ionian natural scientists tried to derive the variety of things from one or a few basic stuffs which produce complexity by mixing and separating in accordance with natural patterns of motion. Plato, to the contrary, reintroduced *value* into the foundation of things by proclaiming reality to be a hierarchy culminating in the Form of the Good: all is, literally, for the best.

Thus, although Plato made no serious attempt to rehabilitate the supernatural beings of pre-Ionian Greek high belief, and indeed shared Xenophanes' prudish disdain for their human-all-too-human foibles, his philosophy is one of reinstatement of high belief. Reason is talked up, but in the end it is the intuition of the authority (the Philosopher) that is to establish the organizing principles and values of the shared worldview. To a great extent Plato's viewpoint has prevailed in the subsequent history of thought, giving high beliefs a new lease on life.

However, let us again recall what reason is, namely, reliable method for finding the truth. But the truth about how things are is expressed only in low beliefs. Hence a "reason," such as Plato's, detached from low beliefs, is a contradiction in terms.

Fortunately, Plato did not succeed in arresting the development of Greek science. Indeed, his school educated and employed the greatest scientist (after Thales) of all time, Aristotle.

THE SOCRATIC SUCCESSION

For better or for worse, there is much truth in the famous remark of Alfred North Whitehead that philosophy is nothing but a series of footnotes to Plato. But there would have been no Platonic text to footnote had Plato not hung around Socrates. Such was the fabulous charisma of the man that after his death, besides Plato at least three other habitués of his circle founded their own schools. Euclid (not the geometer) of Megara emphasized logical studies, but no account of his teaching has survived. Antisthenes, an Athenian but not a citizen, selected the self-sufficiency of Socrates as the core teaching of Cynicism, which evolved into Stoicism, a major philosophy of which more later. And Aristippus of Cyrene (in north Africa), taking off from the Master's possible endorsement of pleasure as the goal of life (in the dialogue *Protagoras*), went home and taught a kind of Hedonism according to which the wise man, freed from conventional hypocrisies, will endeavor to maximize all pleasures, with preference to the most intense, which are associated with eating, drinking, and sex.

Aristippus made his daughter Arete ("Virtue") his intellectual heiress. Her son, also named Aristippus and called "the mother-taught," carried on the school and contributed the teaching that makes it of particular interest in modern times. He asked himself what pleasures are, and answered that they are bodily feelings—which are the only things we can be sure about, for we know them directly, while our knowledge of their causes are inferential and liable to error. Thus when looking at snow, it is better to say not "Snow is white" but "I am being whitened"; better still, "I am being affected whitely." And if someone else looking in the same direction also says this, we can not validly infer that he is having an experience like ours; its inner quality for all we can tell might instead be like ours when we look at grass. This is then the first, and only, appearance in Greek philosophy of a radical subjectivism, an inside-out philosophy. {See Tsouna 1998.} It was offered, however, in the service of the central Hedonism: our choices, as Socrates would insist, ought to be based on knowledge, not mere opinion; Knowledge is limited to feeling; therefore reason demands that choices be based on pleasure and pain, the salient qualities of feeling.

Hegesias, a later Cyrenaic called "the Death-Persuader," anticipated Schopenhauer. From hedonism, surprisingly but straightforwardly, he deduced an unmitigated pessimism. As things are, pains so predominate over pleasures that a life with a positive net balance of pleasure is virtually impossible. Therefore, the rational thing to do is to kill yourself. It is said that his lectures on this theme induced so many suicides that they were forbidden by the government. Hegesias, however, did not kill himself.

Later Cyrenaics took a softer stance on the preference for bodily over mental pleasures, leading to an amalgamation of the Cyrenaic and Epicurean philosophies.

CYNICISM AND STOICISM

Antisthenes, doubtless at least in part because of discriminations he suffered as a metic—a resident of Athens bearing all the burdens of citizenship but none of its privileges—made much of the distinction between what is "by nature" and what is only "by convention," admiring the self-sufficiency of Socrates and advocating a life that would follow nature and ignore the conventional, in a way, a dog's life. (Though *kunikos* means "doggy," the name of the school probably came originally from its meeting in the gymnasium open to metics, the *Kunosarges*.) Cynics tried to outdo each other in eliminating non-essentials, wearing in all weathers a poncho that quickly became dirty, and carrying around a bowl from which they ate their meager meals. This mode of life culminated in the famous Diogenes, who lived in an overturned tub, masturbated publicly, and when asked by Alexander the Great, no less, what he could do for him, replied "Stand out of the sunlight." The Cynics, that is, were classical hippies. Cynicism was more an antiphilosophy than a philosophy.

Nevertheless, when Zeno (of Citium in Cyprus, not Elea) came to Athens and, reading in a bookstore of Socrates, asked whether anyone still around was like him, he was told to "follow that man"—Crates the Cynic, who happened to be passing. He did, and founded Stoicism by downplaying the bizarre behavior and reinstituting Socrates' respect for the pursuit of knowledge.

Stoicism, named for the *Stoa*, "Porch," in the Athenian marketplace where they met, was the dominant philosophy of the Roman Empire until the Christian takeover. Yet completed works by only three Stoics have survived: by Epictetus (more accurately, notes of his lectures taken by the historian Arrian), a lame one-time slave; by the dramatist Seneca, tutor and victim of the Emperor Nero; and by the Emperor Marcus Aurelius. All are primarily ethical in content. But numerous fragments show that Stoics were eminent in various sciences, jurisprudence, and logic. They invented the propositional calculus, international law, and the standard grammatical categories still in use today. And they had a Grand Theory: a materialism holding that the universe is essentially a vast animal, with a world-soul ("Zeus") dispensing Fate wisely and justly (the macrocosm/microcosm idea, going back at least to Anaximenes). Stoicism thus existed on the borderline between philosophy and religion, occasionally falling over it—for example, in accepting astrology (an importation from the Near East) and interpreting omens. The loftiness yet practicality of its ethical teaching, however, remains impressive and influential to this day.

SUMMARY

Frustrated by events from pursuing an active political career, Plato—after Parmenides the first *philosopher* as distinguished from scientist—founded the Academy, a school for educating statesmen. The training was meant to inculcate genuine Knowledge; hence, the curriculum went heavy on mathematics.

Plato supposed the objects of mathematics to be figures and numbers, but not the particular visible marks we deal with—rather, eternal and unchanging things seen only with the mind's eye, which scratches in sand or on wax tablets only "participate" in. He generalized the view into a theory according to which there is such an eternal object—an Idea—for every meaningful word; for to mean, he supposed, is to name. Thus Plato was the first to have a theory of language—on which he built his metaphysical system. It rejected all three Milesian requirements: Unity, Immanence, and Sufficient Reason. Indeed, belief in every aspect of Platonism, save its mathematical and logical content, is necessarily high belief, because the notion of *coping* as test of reality has no place in it.

Plato's metaphysics led to the invention of theology, not a science but a mimicry of one. Though many great philosophers have tried to "reconcile science and religion," the task is hopeless.

Besides Plato, at least three other members of the Socratic circle founded their own philosophies: Euclid a school of logic, Aristippus the Cyrenaic school of hedonism (also the first and only radically subjective Greek theory of knowledge), and Antisthenes, Cynicism which led to Stoicism, the dominant philosophy of the earlier Roman Empire.

CHAPTER 14

༄

Athens III: Aristotle

Aristotle, who in the Middle Ages was with good reason referred to simply as The Philosopher, came from the northern Greek town of Stagira (which had a mostly Ionian population) to Athens and enrolled in the Academy when Plato was sixty years old. He remained in that institution, first as student then as an assistant professor, so to speak, for the rest of Plato's life, about twenty years. His personal relations with Plato seem always to have been cordial; he referred to him after his death as "a man whom it would be blasphemy for bad people even to praise," and spoke often of "our friends the Platonists." Fortunately for us, however, he was no mere disciple. There are hints of fundamental doctrinal disagreements between the two, and it appears that Aristotle thought that Plato in old age had fallen into a sort of number mysticism. At any rate the philosophy that we have from the Stagirite is, on the whole, Ionian and commonsensical, though often with an admixture of Platonic remnants.

Aristotle's written output, even vaster than Plato's, shows differing and even opposite interests from those of the Academician. Plato had nothing to say about the natural sciences except astronomy, which he treated as really a branch of mathematics, whereas (leaving out of account Thales, the first to have the very idea of a natural science) Aristotle was by common consent the greatest scientist who ever lived—preeminent in all and founder of many, most importantly, of biology. The single intellectual field to which he made no important contribution was mathematics!

Plato bequeathed headship of the Academy to his nephew Speusippus, not an impressive figure. Upon Plato's death Aristotle left Athens, going with his colleague Xenocrates to the island of Lesbos in Ionia, where for about twelve years he carried on research mainly in marine biology. For example, he distinguished vertebrates from invertebrates, fish from marine mammals, and made a discovery about the reproductive apparatus of certain cephalopods that was confirmed only two thousand years later. He left Ionia on being summoned to Pella, capital of Macedonia, to be tutor to the crown prince Alexander. Soon afterward his pupil ascended the throne and took off on the conquests that made him The Great, including in his entourage many scientists whom Aristotle recommended, among them his nephew Callicrates. Aristotle returned to Athens and set up his own school, the Lyceum, with financial aid from Alexander. This was essentially a research institute, in the modern sense: it had its own zoo, for example.

In philosophy Aristotle's most lasting achievement was the invention and bringing to completion from scratch of a system of formal logic, the class calculus or theory of the Syllogism, which was the whole of the subject until quite recent times. In its presentation he used letters of the alphabet to stand for classes: the first symbolism.

DIGRESSION: FORMAL LOGIC AND ITS LIMITATIONS

As John Locke remarked, "God was not so niggardly as to make man barely a two-legged animal, and leave it to Aristotle to make him rational." There was logic, of the intuitive sort, before Aristotle's magnificent codification of the class calculus. But this formal system, like all such, takes for granted that all the terms it uses or allows as substitutions for variables are *concepts*, terms having known and finite necessary and sufficient conditions for applicability. No fuzziness allowed. But ordinary discourse is inescapably fuzzy. Vagueness and ambiguity pervade it. For this reason care must be taken in applying the Law of Excluded Middle, that every proposition is either true or false and not both. I once had a run-in with a Fundamentalist preacher (with a Ph.D. degree in Philosophy!) who argued, quite seriously and triumphantly, thus: Every baby is either human or not. So, if Evolution is true, there must have been at least one human baby born of an ape (non-human) mother. But that is absurd. Therefore logic proves that Evolution is false. {See Warren and Matson 1978, 247.}

Nor is the propositional calculus (an invention of Stoic philosophers) free of limitations. The truth-table definition of truth inherits the difficulties of the Aristotelian syllogism, while the definitions of conjunction, alternation, and implication are grossly inadequate as substitutes for these notions as they function in ordinary discourse. So much the worse for ordinary language, say some philosophers, and invent artificially more "precise" languages to be more suitable for science. But typically these turn out to be incapable of even stating what needs to be said in the natural sciences. For example, the following simple argument

Whatever X believes, Y believes;
X believes that *p*;
Therefore, Y believes that *p*.

cannot be stated in the usual unsupplemented notation. {See Bealer 1982, 42.} Fixing them up with more and more complicated and ad hoc modifications seems an enterprise subject to a law of diminishing returns. The working scientists themselves seem to be satisfied to get along on their intuitions— which in the end is what the logicians themselves were doing, really.

The aim of all sciences, Aristotle declared, is to advance from the facts, given in perception and stored in the memory, to the *reasoned* facts, *necessarily* true accounts of why these facts *cannot be otherwise*.

> We say that that which cannot be otherwise is necessarily so. And from this sense of necessary all the others are somehow derived ....Demonstration is a necessary thing, because the conclusion cannot be otherwise, if there has been demonstration in the full sense; and the causes of this necessity are the first premises, i.e. the fact that the propositions from which the deduction proceeds cannot be otherwise. (*Metaphysics* Book 5, Chapter 5, 1015a33.)

He describes this process of demonstration thus:

> All animals.... have a connate discriminatory capacity, which is called perception. And if perception is present in them, in some animals retention of the percept comes about, but in others it does not come about. Now for those in which it does not come about, there is no knowledge outside perceiving.... ; but for some perceivers, it is possible to grasp it in their minds. And when many such things come about, then a difference comes about, so that some come to have an account from the retention of such things, and others do not.
>   So from perception there comes memory,...,and from memory (when it occurs often in connection with the same thing), experience; for memories that are many in number form a single experience. And from experience, or from the whole universal that has come to rest in the soul,...there comes a principle...of skill if it deals with how things come about, of understanding if it deals with what is the case. (*Posterior Analytics* 100a3–9, trans. Jonathan Barnes.)

"Universal" in this passage—for example, "man" as opposed to "Socrates"—is what has become of the Platonic Idea: Aristotle has put it back into things, as their Nature or Essence. And as against the naive view of Plato, that every intelligible word must be the name of some individual entity—hence the separated Ideas—Aristotle in his *Categories* develops the implications of his insight that "Being is said in many ways": of Substances, the individual things that exist in their own right, then of quantity, quality, relation, place, time, posture, action, and passion, all of which are either "in a subject" or "said of a subject." What there is, really, is this man—tall, married, standing, and so on; this horse—piebald, whinnying, and so on. All men are (perhaps) rational; what this means is that they have a common property, and have it essentially. Aristotle went further in distancing himself from the Ideas, pointing out that they duplicate the world without explaining it—they don't *do* anything, and they give rise to gratuitous puzzles about the nature of "participation."

Plato considered change, of any sort, to be fundamentally irrational: If A becomes B, then something that *was* true no longer *is* true—an intolerable affront to Reality. Aristotle, on the other hand, abstracting from his primary interest in biology, sought to analyze change in terms of potentiality and actuality: the oak is the actuality of which the acorn is the potentiality. This can be further explained as the acorn's being the matter which (together with air, soil, and water—more matter) naturally takes on, bit by bit, the form of the oak. There is no affront to reason anywhere in the process.

A NOTE ON ESTABLISHING NATURAL NECESSITY

Prior to literacy, we may already distinguish two kinds of words. The first are *labels*, pointers. "Hesperus" is an example. Outside at dusk we point to a bright object in the sky and we say—baptize it, as it were—"That's Hesperus, the Evening Star." Or we say "That's water" of the stuff that we haul out of the well and drink. Not just nouns, but verbs "He's walking," "He's running," "He's sleeping;" adjectives—"That's heliotrope;" adverbs—"He did it quickly;" abstractions—"It's probably going to rain tomorrow," "the witch's hex caused the milk to sour." Also in common use are *concepts*, words that do have explicit definitions that (nearly) everybody can recognize: uncle; second cousin; numbers; net worth; husband; Pope; miles-per-hour.

With literacy come dictionaries, which are books that try to turn labels into concepts by specifying necessary and sufficient conditions for their proper use. But they do not always succeed in doing so. What are the separately necessary and jointly sufficient conditions for being a house? Or bald? Or a pig? Or a cause?

More important, progress in science often results in sharpening and revising concepts. For instance, "whale," beginning as the label for a big denizen of the water that spouts, is transferred from the class of fish to that of mammals. And "sulfur," label for a yellow powder with acrid scent when burned, becomes "Element No. 16," a description that embeds it in the highly complex *theory* summed up in the Periodic Table of the Elements as that one whose nucleus contains just sixteen protons.

A properly formed concept does have necessary and sufficient conditions for sorting the world into those items that fall within its extension, and those that do not. Consequently concepts can enter into rigorous logical relations of entailment with one another. Labels as such cannot do this. Therefore the terms of a proper *theory* should all be concepts. A theory, then, like a geometrical proof (which indeed *is* a theory) will, in a priori fashion, deduce consequences from premises constituted from concepts. It will state, that is, that such and such a conclusion *necessarily* follows from the concepts in the premises. This will be so whether or not the concepts involved match up with the associated labels. But if they do in fact so match, then the merely hypothetical necessity of the theory transfers to the factual necessity of the theory's subject matter—in other words, *necessary truth about the nature of things.*

A simple example: Is it a necessary truth ("could not be otherwise") that wet clothes hung out in sunlight will dry? Yes. The concept "H_2O" is that of rapidly moving molecules which below a threshold value of energy exist in the liquid phase, in the gaseous phase when above it. The concept "sunlight" is that of photons, carriers of energy which transfer it to what they strike. Hence, when enough of them strike the H_2O molecules the latter will enter the gaseous phase (which happening is what drying out *is*).

But let us be careful. Of course the clothes will not dry out if a rainstorm comes up, or if they fall off the line, and so on. But these contingencies are irrelevant. What is more worrisome is that the label/concept identifications could be false. Granted: Priestley, Lavoisier, and all those people might have been wrong. But that only requires us to say that the identifications, *if true at all*, are necessarily true. There is no such thing as "contingent identity." Those philosophers who say there is, argue that there are "possible worlds" in which stuff just like water here on earth to all empirical inspections is nevertheless XYZ instead of H_2O. Or that there might be (for all we know) a law of nature that after a certain date the dew point of H_2O will increase tenfold. In chapter 18 these worries will be shown to be survivals of the medieval metaphysic of the contingency of the world, for which there is no warrant in scientific theory or practice.

For a more extended treatment of this subject see {Matson 1976, chapter 2.}

FINAL CAUSES

All this was an immensely salutary putting of philosophy back on its Ionian track: restoration of Unity, Immanence, and *Logos*. The process was not quite complete, however. Aristotle prided himself on what he claimed as his own invention, the Final Cause. To explain anything, he averred, you must answer four questions: What is it? (The Formal Cause) What stuff composes it? (The Material Cause) How did it come into existence? (The Efficient Cause) and What is its For-the-sake-of? (The Final Cause). That is, the Philosopher believed that everything has a *telos*, an end, a what-for. This seemed to him to follow from his analysis of change: the actuality is the *goal* of the potentiality—a very natural and convincing way to look at things, particularly in biology. But it is bringing back teleology, an explanatory category that Thales had rejected in founding science; and not exactly by the back door either. For not only does every individual thing have its *telos* (either as end in itself or as means to another end) but so does the whole Cosmos. What makes the world go around? God, Aristotle said; the Unmoved Mover; Thought thinking of Thought (Parmenides!) moving the celestial spheres "as the beloved moves the lover." Good poetry, perhaps, but more what one might expect of Plato.

Theophrastus was a worthy successor of Aristotle as head of the Lyceum, but the school did not survive him. Studies of the type that had gone on there migrated to the Museum of Alexandria in Egypt, a city at a mouth of the Nile that the Great one had founded. A sad consequence was that the Lyceum library was broken up, with loss of an estimated two-thirds of Aristotle's writings.

A NOTE ON INFERRED ENTITIES

Inferred entities are as old as imagination, which is forever creating them. Gods, spirits, souls, mana, and so on are imaginative creations that "explain" the facts of experience. So is Anaximander's "Boundless" and his shark-like ancestors of human beings. But there is a difference.

Aristotle (*On the Heavens* II 14, 297a8–298a20) made these observations: (i) When the moon is eclipsed (which, he correctly believed, happens when the earth passes between sun and moon) the shadow moving across the moon is curved. (ii) When one travels north or south (but not east or west), stars alter their relations to the horizon. And he described a "thought experiment" (iii): If the earth came into being from loose parts, each with an impulse to move to the center of the world, they would jostle one another so that eventually the whole would be spherical. Putting these three points together he concluded that the earth is "necessarily spherical."

Was his belief high or low?

High, by our definition, as it was not a belief that when acted on was found not to occasion surprise, for (at the time of its formulation) it had not been acted on at all; nor did the low beliefs that Aristotle cited logically imply it. But it was a high belief of a very different sort from belief that Zeus hurls the thunderbolt, and from Plato's beliefs about Ideas. The difference lies in its close relationship to low beliefs, its eligibility, so to speak, to become a low belief. (Tethered beliefs are analogous to the "verifiable propositions" of the Logical Positivists.) If high beliefs are balloons, those generated from stories are free-floating, whereas scientific hypotheses are *tethered.* They can be pulled down to earth.

Two of the "tethers" on Aristotle's hypothesis of a spherical earth are the observations of the curved shadow of the earth and the displacement of the horizon with travel north or south. One end of each "tether" is "on the ground," the other attached to the hypothesis in this sense: they confirm it, that is to say, if the earth is spherical, then these facts are just what would be expected—indeed, what *must* be observed. Further, it is hard to conceive how these facts could be as they are if the earth had any shape other than spherical. If we could establish that no other shape would be compatible with the observations, then indeed we would have a logical tether and would be justified in counting the sphericity of the earth as a low belief. But it seems that a drum-shaped earth, with us on the curved portion, is not absolutely ruled out by these two items.

The thought experiment (how the earth might coalesce from separate parts) likewise is tethered to some low beliefs about the way things behave; but its main attachment is to certain other Aristotelian theories, in particular that of "natural place," making the logical situation too complicated for discussion here.

We may note that the three considerations that Aristotle adduces interlock and thereby reinforce one another, as in a crossword puzzle. The Zeus theory, on the other hand, is completely untethered—there are no low beliefs about Zeus. Nor does it explain the lightning. Since Zeus can do almost anything (not absolutely anything; Zeus is subject to Fate), to say "Zeus willed it" gives us no information about what it was that happened, nor why. A personal worldview may "make us less forlorn," but it tells us nothing of the nature of things, does not orient us in the world.

Tethered high beliefs are (what are ordinarily called) scientific hypotheses, if they are beliefs at all. Scientific protocol nowadays regards it as a delict for a scientist to actually believe a hypothesis, especially his own, before it has passed all the tests and thus become a low belief. Few ancient scientists exercised any such restraint, or even seem to have been aware of the possibility of merely entertaining a proposition. Epicurus, typically, propounded numerous explanations—of lightning, for instance—and recommended that one believe all of them simultaneously! This is not the place to go into the question of how much tethering makes a belief justifiable. Ultimately the requirement is that the belief be part of a network or system that "works," such as the Periodic Table of the Elements or quantum physics.

Our senses are given to us so that we can get our dinner. And so that we can fly through the air, and to the Moon. There is no difference in principle.

It is in these terms, too, that we may reply to the objection to evolutionary epistemology, that no such account can explain how it has come about that mankind has been able to develop objective understanding of the principles by which the universe works {see, e.g., Nagel 1986, 78–82}, the "true theory of the universe." For the circumstance of believing or disbelieving or never having heard of the true theory of the universe cannot have any bearing on whether one is biologically successful, leaves descendants—which is the nub of every evolutionary explanation.

If the true theory of the universe were a high belief, the objection would be valid, for evolutionary force cannot explain why any particular high belief exists and is held. The most it can do is show how a propensity to accept kinds of high beliefs that have certain consequences such as social harmony and cohesion would be selected for. And there does not appear to be any reason why the true theory of the universe should confer any such advantage on those privy to it.

However, the true theory of the universe is a low belief, linked by logic and mathematics to paradigm low beliefs, and is itself a serendipitous consequence of the ability to form high beliefs. An outline sketch of the biological development leading to it is this. Multicellular animals lead to perception; perception leads to low beliefs; believing (as opposed to mere instinct) as a means of coping leads to big brains; big brains lead to language; language leads to imagination; imagination leads to high beliefs, and also operates on low beliefs, leading to science-systematized low beliefs). Science (without further help from biological evolution) leads finally to the true theory of the universe, which, then, is a by-product—unintended, to be sure, but of course evolution doesn't intend anything–of more and more complicated and ingenious stratagems for getting one's dinner, finding one's mate, and staving off one's enemies.

SUMMARY

Aristotle, next to Thales the greatest scientist of all time, was for twenty years student then junior colleague of Plato in the Academy. He then spent twelve years in and around Lesbos doing mostly marine biology, after which he became tutor to Alexander, crown prince of Macedonia. When the latter went off on his conquests, Aristotle returned to Athens, where he founded his own school, the Lyceum, essentially a research institute.

Aristotle was eminent in, very often founder of, every intellectual endeavor of his time, except mathematics. For example, he was the first formal logician, bringing the class calculus to completion from scratch. His conception of natural science was that of starting from facts of observation, to proceed to the reasoned fact: explanation of the data by necessary truths, which could not be otherwise. His account of how language works pulled the rug out from under Plato's basis for postulating Ideas. The Aristotelian universe therefore has only one story. Reality is basically the sum of individual substances—this man, that horse—of which qualities, relations, and so on are predicated. As against Plato, who thought all change to be irrational, he analyzed movement into progression from being potential to being actual, matter taking on form. His biological outlook, however, led to his insistence on Final Causes, a reintroduction into philosophy of the teleology that Thales had banished in the beginning.

Inferred entities enter into both scientific and pre-scientific explanations, but with a difference. The inferred entities of science are "tethered," i.e. hypotheses that are candidates for low-beliefhood.

The objection to Evolutionary Epistemology that it cannot explain how we (may) have arrived at the true theory of the universe, because neither belief nor disbelief nor ignorance of it can have survival value, is met by pointing out that the true theory of the universe is the serendipitous consequence of progressively more complicated and efficient means of coping.

CHAPTER 15

⟡

Alexandria

After Alexander's death his generals fought each other for supremacy. At the end of the fourth century B.C, his empire had effectively been divided into three: the Macedonian, including all of Greece; the Seleucid, roughly his eastern conquests; and Egypt. The last was ruled by Ptolemy I Soter ("Savior"), who took the title Pharaoh. He was a great patron of Greek culture, founding in the new city of Alexandria the Library, which grew to an institution of perhaps more than a million volumes, and the Museum (temple of the Muses), the greatest scientific research institute of antiquity.

Science thus governmentally subsidized made great strides. The astronomer Eratosthenes calculated the size of the earth almost exactly. Another astronomer, Aristarchus, put forth the theory, which did not win acceptance, that the earth revolved around the Sun, not vice versa. Mechanical devices were built including pipe organs, calculators, and a steam turbine (which, however, found no practical use).

Ancient mathematics culminated in the work of Archimedes, who was born (287 B.C.) and lived all his life in the Sicilian city of Syracuse but was, so to speak, a "corresponding member" of the Museum. He proved, for example, that both surface area and volume of a sphere are two-thirds those of a circumscribed cylinder.

In philosophy and literature it seemed that the Greek genius was spent. The Library was the province of scholarship, producing excellent editions of the classics but no important new literature. In philosophy the schools of Athens retained their preeminence, but they too became professorial and uncreative. If there can be said to be an exception to this generalization, it lies in the rise of Skepticism—at first, oddly, in the Platonic Academy!

GREEK SKEPTICISM

The history and significance of skepticism need to be reconsidered in light of the distinction between high and low beliefs.

The word "skeptic" is from the Greek for "inquire" or "investigate," "scrutinize," not "doubt." Socrates often said "Skepteon," "this ought to be investigated," when someone brought up an interesting thesis. To be sure, Plato represents him as usually coming up with a negative conclusion; hence perhaps the shift in meaning, and the "skepticism" of the Middle Academy. The Skeptic is (or at first was) one who looked into the bases of beliefs that other people swallowed whole in accordance with the built-in propensity of language users to believe what they are told. Science itself began as skepticism in this sense, and still is, or ought to be, thus skeptical.

Of course there is a natural connection between inquiry and doubt. John Dewey held that doubt, an irritating state of tension, is the motive for all inquiry, the purpose of which is to remove it. But this generalization, while accounting for many investigations, criminal and financial, for instance, is too broad; as often as not, scientific inquiry creates more doubts than it alleviates, since it brings into question the received opinions. We have seen how this came about on the grand scale in fifth-century Athens.

Moreover, doubt is not exactly the right word for the state of mind induced by contemplation of scientific discoveries. Socrates in *The Clouds* did not doubt that Zeus produced the rain; he rejected that notion entirely. Likewise Xenophanes with anthropomorphic divinities. And the nineteenth-century clergyman who, having read Lyell and Darwin, began to "have doubts," really was lost already—he was treating his religion as a hypothesis, not a faith.

The terminology of high and low beliefs makes possible a simple description of Greek skepticism, which some historians have misunderstood as belonging to the same genre as the doubts associated with Descartes. Greek skepticism, or Pyrrhonism (after Pyrrho, who went with Alexander's troops to India where he met the Gymnosophists, "naked philosophers," presumably Brahmin sages) was simply the philosophy that utterly rejected high beliefs, whether tethered or not. "It appears to us," Sextus Empiricus (a skeptical writer of late antiquity, whose works have survived) wrote,

> that honey is sweet. This we concede, for we experience sweetness through sensation. We doubt, however, whether it is sweet by reason of its essence, which is not a question of the appearance, but of that which is asserted of the appearance.

Pyrrhonists never doubted "the external world"; they doubted whether anything is or can be known about its "inner nature." Their arguments recapitulate the considerations we advanced when dealing with sense perception, that this encounter with the world is fundamentally a means of coping, the production of an internal model for action, which can be known to be "true" only in the sense that it does not misguide our practical endeavors. Being our sizing up of the data of sense, it contains uneliminably subjective elements. This is what the Pyrrhonists meant by "appearances." The "essence," on the other hand, was whatever was supposed to explain why the thing appeared as it did: the spirit inside, or the god outside, or the Platonic Form, or the Aristotelian Nature, or atoms, or what have you. Belief in any of these things was high belief, and the Pyrrhonists would have none of it. The kind of argumentative inference that we have called "tethering" was rejected. Although the Pyrrhonists did not reject logic—which is, as we have pointed out, low—they denied that it could establish the reality of an inferred entity.

To every argument, they held, another of equal weight can be opposed: one, namely, from opposite assumptions.

The Pyrrhonists—ancestors not of modern Skeptics, but of Positivists and Pragmatists—were right for their time. Ancient science was strongest in mathematics, where the necessity is, so to speak, right on the surface; and in taxonomical endeavors, such as Aristotle's magnificent biology. However, with few exceptions, the most notable of course being atomism, its theorizing missed the mark. Its inferred entities, especially those supposed to be causal, are seen in hindsight to have been of the wrong types. So the Pyrrhonist claim, that no high belief is reasonable, was (almost) correct.

Nevertheless, it would have been catastrophic had their view prevailed. They could not know that the imaginative element in science was capable of being disciplined and that it would eventually become productive of real truths involving inferred entities.

SUMMARY

Ptolemy, one of Alexander's generals, a patron of Greek culture, founded in Alexandria, capital of his Egyptian empire, the Library, the greatest depository of Greek literature, and the Museum, a research institute. Science made great advances in the Museum. The Library produced great scholars but little new literature.

The principal philosophical innovation of the period was the rise of Skepticism, which utterly rejected high beliefs, whether tethered or not. Greek Skepticism is the ancestor of modern Positivism and Pragmatism, not of Cartesian skepticism. It was quite correct for its time, but it is a good thing that it did not prevail, for it would have eliminated the element of imagination that is essential to science.

CHAPTER 16

✿

Beliefs About Believers

Hitherto we have been discussing beliefs about the world at large, before and after the invention of science in Miletus. Now it is time to say something about the beliefs about themselves—what human beings are—held by the people who do the believing.

All humans are *at least* bodies. Both Egyptians and Greeks developed advanced surgical technologies—they could, for example, trepan and remove cataracts. Greek medicine has some claim to have been the first science; it was empirical, that is, based on careful observation and thus on low beliefs; and early on it eschewed involvement of supposed divinities and was skeptical of sweeping theories. Naturally, practitioners of this medicine came to acquire considerable anatomical knowledge.

But people everywhere tend to divide their conceptions of themselves into two: their bodies, and something else. The something-else, a high-belief imaginary entity, like such constructions generally, has different characteristics in different cultures. In Egypt, which was obsessed by the notion, it was the *ka,* a sort of inner duplicate of the body, which survived death, and being what one is good or bad with, might have a happy afterlife or the reverse after judgment by the god Osiris. If you seek its monument, look at the Pyramids.

In early Greek thought, however, the something-else is far less robust. At the very beginning of the first work of European literature, the *Iliad,* Homer refers to the dichotomy, singing that the destructive wrath of Achilles sent down to Hades the souls (*psuchai*) of myriad Achaeans, but left *themselves* to be dainties for dogs and feasts for birds. That was to say, Homer thought the body to be primarily the person, the soul only something that accounted for the difference between a live hero and his corpse: the life principle, we may vaguely call it. Since the main difference is that the hero breathes and bleeds, and the corpse doesn't, Homer's culture discerned two souls or soul parts: the breath (*psyche* is an onomatopoetic word; so is *spirit*) and the blood. In the *Odyssey*, where post-mortem souls are visited in the underworld, they are wraiths (breaths) that flit about, unperceiving and irrational, capable of communicating about their miserable lot only if given a drink of blood. All the same, they are supernatural beings and live (if you call that living) forever.

Thales, as we have seen, mentioned the *psyche*, saying that the stone of Magnesia has one because it moves iron. We interpreted this remark as implying not a vivification of a rock but a demystification of the life principle; it becomes just whatever is a source of internal motion. Anaximander, probably to his credit, had nothing to say about souls, as far as we know. The notion, however, recurs in Anaximenes, in a full-blown form. Being first to express the *macrocosm/microcosm* idea that the human being is a small-scale copy of the whole cosmos, he declared that the soul is Air (or Mist), holding together the body as the Air out there holds together the universe. Aristophanes made fun of this doctrine in *The Clouds* when he had Socrates enter suspended in a basket and declaiming about "subtly blending his intellect with its kindred air." For implicit in this view is the notion that we have found in Parmenides, that the whole universe is a thinking as well as an extended being. But if so, it seems then that at the death of the individual person the *psyche* is simply dispersed into the rest of the Air, retaining no individuality.

But Pythagoras and his follower Empedocles reversed the Homeric priority of body to soul. Like the Egyptians, they conceived the soul as a unity that is the real person, distinct from and surviving bodily death; moreover, it is capable of inhabiting other bodies, not necessarily human, in succession: transmigration, metempsychosis. (Where they got this notion is unknown; Scythia and India are suspected.) Pythagoras claimed to remember having been a trooper in the Trojan war, and Empedocles sang of having been "a boy and a girl and a bush and a bird and a dumb saltwater fish." This crazy teaching was, however, of the utmost consequence for science and philosophy, since it was for Pythagoreans the principal motivation for engaging in these activities, and the reason they formed themselves into a Brotherhood with a life of its own. Successive incarnations, they held, were not at random, but depended on the stage of purification so far reached. Further, the series of incarnations might be brought to a happy ending in a permanent spiritual state by diligent care of the soul. These are essentially the doctrines of *karma* and *nirvana*, hence, a strong reason for suspecting an Indian origin. But it was typically and touchingly Ionian Greek to take it as obvious that clear thinking about mathematics, for example, would have this purifying effect. Socrates and Plato, who may have been Pythagoreans and certainly were strongly influenced by them, believed in some such version of immortality.

NOUS

Homer recognizes a faculty of the human being, presumably associated with the blood-soul, *nous*. "Mind" is the usual translation, but intelligence, recognition, judgment, apperception, intuition, sizing-up, all are needed to convey its particular sense. The verb *noein*, to size up, is of more frequent occurrence than the noun. The word comes to the fore in the poem of Parmenides, where it is said, among other things, to be "the same" as Being. As we have seen, Parmenides believed not only that basic reality was the Space of geometry, but that this reality, which is what real grade A thought is about, thinks back, so to speak. This was because Parmenides supposed, like many other philosophers, that knower and known must share something in common; and what could have anything in common with Thought except more Thought? (A form of this idea recurs in Aristotle's description of God's activity as "thought thinking of thought.")

It remained for Anaxagoras, however, to make an impersonal Mind into the major cosmic force. What Mind—any mind—essentially does, he noted, is bring order into things. The Cosmos is an ordered thing—that is what the word means. But things in the beginning, we recall, were completely mixed up—all except Mind, which could retain its power because it was the only thing that existed all by itself, a kind of seed, it seems, but finer and subtler than the blood-seeds. So Mind was able to start and guide the rotation that in the common style of Ionian cosmologies separated things out to a certain extent. And since the ordering of Mind is not merely any old ordering, but goal-directed, it was all for the best. Thus Anaxagoras reintroduced teleology into cosmology, a reversion, as we have noted, to pre-Thalesian thinking. It pleased Socrates when he heard of it, but as he read further in Anaxagoras's book he was disappointed to find that the author explained all that he could in mechanistic terms, dragging in Mind only as a gimmick when all else failed. So Socrates and Plato resolved to be more consistent in their employment of this principle.

Plato was the first philosopher to have a detailed theory of the soul: the famous tripartite entity worked out in *The Republic*, composed of Reason (*dianoia*, essentially the same word as *nous*), Spirited-element (or however one is to translate *thymos*: "Will" is not quite right), and Appetite. They dwell in the head, the chest, and farther south, respectively. (The parallel with Freudian Superego, Ego, and Id is striking but probably accidental.) Plato explains them by analogy with the three classes of the Ideal State, which should be ruled by philosophers who with the aid of the armed forces keep the civilian laity in line. The qualities of these soul elements are innate and unequal as between individuals. Only those with the highest IQs (Plato would certainly have used these tests and others in the battery of the modern psychologist had he had them) are given the rigorous graduate education necessary to make them ready to intuit the Ideas, culminating in acquaintance with the Idea of the Good and thus capable of directing the State. There would be no need for hierarchy among the philosophical rulers, for Reason is the same in and for all. (Compare Rawls's contractors behind the "veil of ignorance," i.e., elimination of all bias-inducing personal knowledge, who would, he assures us, unanimously adopt his proposed constitution guaranteeing Justice.) Moreover, it is this Reason that is immortal. Plato argued that the soul must be immortal because, being immaterial, it is without separable parts and hence exempt from destruction. Being aware that in that case it must exist before birth as well as after death, he turned this feature into a theoretical advantage: since we know Absolute Justice and Absolute Equality without ever having seen any two things that were absolutely equal, or having experienced any arrangement that was absolutely just, it must have been in its pre-natal state that the soul became acquainted with these ideas. (Just how, we aren't told.) Further, the soul being not an idea but capable of knowing the ideas, it must have something in common with them, namely, their eternity.

Indeed, Plato famously described the human being as a soul *imprisoned* in a body. This conception, modified in some respects, is the principal contribution (if that is the right word) of Greek philosophy to Christian theology, as we shall see.

ARISTOTLE ON THE SOUL

There was a Pythagorean theory, known to us only from Plato's argument attempting to refute it, according to which the soul was not a thing (material or immaterial) at all but an "attunement" of the body—not the body nor an extra organ of the body but its functioning, as music is of the musical instrument. We can see that this fit neatly into Pythagoras's conception of the doing of science as purification. The bad soul is just the one defectively tuned. This conception is more than a mere metaphor, as it transfers the soul and its doings into a completely different category from that of the naive interior double of the Egyptian *ka* and such. It may also have been what suggested to Aristotle his own functionalist conception.

Aristotle's fundamental model for explanation of process, as we have seen, was growth: the development from potentiality to actuality, which he also expressed in terms of matter taking on form. The soul of a living thing, in this scheme, becomes the form, which is the same as the functioning, of the (mature, actual) body. "If the eye were a complete body, seeing would be its soul." There are three kinds or grades of souls: vegetative, animal, and rational. The first is basic vital functioning, the life principle, what makes the difference between a living being (whether stalk of celery or professor of philosophy) and an inanimate object. The animal soul is found not in all living beings but only in those capable of perception and movement. And the rational soul (in effect, though the Philosopher did not put it that way) is the intellect of animals having language: us. So soul becomes a biological concept. The word "three" in its description must not be taken too seriously. Any organism functions as a unity; there is interaction and feedback from rational to animal and vegetative; the basic functioning of the professor is very different from that of celery.

Yet here as elsewhere Aristotle is a mixture of common sense and Platonism. There *is* something special about the rational soul—which turns out, unsurprisingly, to be immortal. The rational soul of the individual is on loan, so to speak, from the cosmic Reason—the thought that God thinks about—and at death it reverts to it; retaining, however, no individuality. So there is no Aristotelian afterlife otherwise than in this highly attenuated sense.

Aristotle's treatise *On the Soul* would have been the foundation of the science of psychology if any Greek beyond his successor Theophrastus had followed up on it. It is the only ancient work on the subject that is largely based on low beliefs. But sadly, it was universally neglected, and the idea of functionalism had to be rediscovered in recent times.

DEMOCRITUS AND EPICURUS

Overall, Atomism was the ancient philosophy that most faithfully preserved and continued the Thalesian triple insight: Unity, Immanence, Reason. But in some respects it failed, as we have noted, to incorporate scientific advances made by other schools, notably those deriving from the Pythagoreans. One of these was the conception of soul. Epicurus and Lucretius still spoke of soul atoms, particularly light and mobile particles, aggregations of which were conscious; no atom by itself, however, bore consciousness, which like color and taste was subjective, "by custom." We can see in hindsight that the atomists would have compromised none of their principles had they adopted the functionalism of Aristotle. Indeed, atomism plus functionalism is roughly the ontological underpinning of scientific psychology in our day.

Soul atoms, however, were very useful to them, from their point of view. We must remember that the main motive behind Epicurus's philosophy was emancipation of mankind from the terrors of an afterlife threatened by priests. Consequently, a reason to *dis*believe in immortality was a primary desideratum, and this the atoms handily provided. At death the soul atoms, like the Air-soul of Anaximenes, simply dispersed into the atmosphere with the last breath of the decedent; and that was that.

SUMMARY

Medicine, like engineering and mathematics, in Egypt and in pre-Thalesian Greece was an empirical art based largely on low beliefs. But beliefs about the something-else besides the body that constituted a person—the "soul"—were high, almost by definition; and in consequence there were very many different soul conceptions, ranging from the Egyptian *ka*, a sort of interior double, to the immaterial but tripartite soul of Plato and the functionalism of Aristotle. Most included some kind of immortality, with a heaven for the good and a hell for the bad. Often these conceptions were bound up with theories of the universe, for example, that of Anaximenes that a person is a small-scale copy of the cosmos, Air being what in both cases "holds them together."

Pythagoras initiated the theory of a soul separate from body, migrating between bodies of different animals and even vegetables. This odd belief, however, motivated his science, which he conceived as a process of "purification" that might at last bring to an end these painful bodily imprisonments.

Mind (*Nous*), associated with but not identical to soul, was the faculty of discerning and imposing order on things. Anaxagoras made it into the cosmic force, thus reintroducing into science and philosophy the teleology that Thales had banished. Plato developed this idea further, and even Aristotle's science is infected by it. However, Aristotle rightly transferred soul from the category of Substance (Thing) to the categories of Action and Passion (Functions). The soul, in his terminology, is the Form of the Body. Simple organisms have a "vegetative soul," the bare life principle; in animals it complicates into the "animal soul" which includes perception; and in us to the "rational soul"—really the "language-soul." This last-named shares in immortality, though not as an individual.

The Atomists retained the concept of soul as a thing, or rather things: there are special soul atoms, lighter and more mobile than others, though consciousness is not a property of an individual atom but something that (like color and taste) supervenes on combinations of soul atoms. Epicurus was thus able to dismiss fears of Hellenistic hells, for at death the combinations disperse, like the Air-soul of Anaximenes.

Alone among ancient soul theories, Aristotle's, being based to some extent on observations (low beliefs), and getting the ontology right (action-passion), may claim a (qualified) scientific status. It was Plato's, however, which survived into Christian thought, being in fact the principal Greek contribution to Christian theology.

PART III

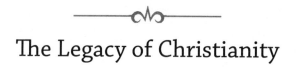

The Legacy of Christianity

CHAPTER 17

⚬⌁⚬

Jerusalem Collides with Athens

THE RELIGIOUS SITUATION IN THE HELLENISTIC WORLD

Perhaps some of what has been said up to this point may have given the impression that after Thales, atheism was rife in the ancient world. But that would be a misunderstanding. Thales himself, remember, said that "all things are full of gods." True, those gods, being the immanent forces of nature, were in many ways unlike how the good people of Miletus conceived the gods that they worshiped and sacrificed to in their temples. But they were "ageless and deathless," in Anaximander's description, that is to say, they met the minimum requirement for divinity. And anyone in Miletus was free to conceive of gods as he or she pleased; the Ionians had no "book," no official theology, and no dogmas. Homer and Hesiod were in no way "bibles;" and Homer himself, at least in the *Iliad*, hardly maintained a reverent attitude. Even the Epicureans affirmed the existence of gods—after all, people *saw* them in their dreams—but said they just didn't take any notice of us; though even so they retained some usefulness as models of The Carefree Life. In all of ancient Greek thought, only two men explicitly denied the very existence of gods: the Cyrenaic philosopher Theodorus (whose name, ironically, means "Gift of God") "the Atheist," and Euhemerus who theorized that gods were simply eminent men of long ago about whom fables were told.

Besides, it must be kept in mind that the science and philosophy we have been discussing were mainly the province of a statistically tiny subclass of the upper classes. Most people had no interest in these matters but just went ahead with their traditional rituals, observances that were more like saluting the flag than being born again. That was why Aristophanes' *Clouds* didn't win the Oscar for 423 B.C.

But to emphasize this widespread attitude to official religion may push one into the opposite error of supposing that there was no such thing as what we would recognize as religious fervor in the ancient world. There was in fact plenty, centering on the so-called Mystery Cults. Some were of ancient and native Greek origin, such as the Eleusinian Mysteries of the Athenians. But most, in Hellenistic times, were importations from the East and from Egypt. All of them required formal initiations, often expensive, in which their secret doctrines were revealed; and the principal teaching in all of them was that the soul of the initiate who behaved properly would attain immortality. (Yes, Pythagoreanism qualified as a mystery, though more intellectual than the others.) They coexisted peacefully, like lodges.

The Romans wisely abstained from interfering with the religions, official and unofficial, of the peoples they conquered, as long as (in Imperial times) they acknowledged the divinity of the Emperor. (With one exception, to their credit: they closed down the Carthaginian temples to Baal in which firstborn babies were incinerated.) The high point of toleration came in the reign of the Emperor Tiberius, who abolished prosecutions for blasphemy, remarking that mistreated gods ought to be able to take care of themselves.

And so it might have gone on indefinitely in this amiable way had it not been for a single mystery sect that refused to conform to these rules: the Christians. Their defiance of Roman policies at length led the government to try to stamp them out. It failed, with consequences fundamentally affecting every aspect of Graeco-Roman culture, particularly philosophy, which the Christians in their turn tried to stamp out, with more success.

THE CHRISTIAN WORLDVIEW

The religion of the Jews was very unlike other Middle Eastern beliefs. It had its Book, the Pentateuch, held to be divinely inspired and therefore totally free of error. Its very first words were these:

In the beginning God created the heavens and the earth.

God, singular. Only one. And the way He did it was simply by issuing commands:

And God said, Let there be light; and there was light.

And so on, through six days, concluding with the creation of the first man and woman. (There is a second account of creation, beginning at *Genesis* 2.4, according to which God fashioned man out of the dust and after literally breathing life into him, anesthetized him, removed a rib, from which He sculpted the first woman. But this version is from an earlier stage of Hebrew thought.)

In its elegant simplicity this story is the ultimate to which the Command side of the human action model can be carried: creation of all things out of nothing at all, by unlimited power with no intermediary whatsoever. In *Genesis* 1.2, "and the spirit of God moved upon the face of the waters," there is a reminiscence of the usual Middle Eastern water cosmogonies; but it was disregarded in subsequent thought.

Unlike all other gods of the Mediterranean basin, this God was both supreme warlord of His people and also supreme legislator, commanding observance both of the simple precepts of the Ten Commandments and the very detailed codes in *Leviticus*. Being a "good" person consisted in obeying them; the single motive for moral behavior was fear of punishment for disobedience—the act that notoriously got the first man and woman, and all their descendants, into so much trouble.

This God not only created all lesser beings but He also laid down the laws of how they were to behave. And He could—and did—suspend those laws when His purposes required. The Sun went regularly around the Earth once a day, except when more daylight was needed for His agent Joshua to finish off the Amorites (*Joshua* 10.12–14). He parted the Red Sea to let the fleeing Israelites through, then joined it again to drown the pursuing Egyptians (*Exodus* 14).

THE TRIUMPH OF CHRISTIANITY

Christianity, a mystery religion that began as a heretical sect of Judaism, inherited these conceptions of God and His relation to His creation. When it captured the state apparatus of the Roman empire and imposed itself by force, it meant extirpation of the Greek conception of the world and revival of the human action model of explanation—but with a difference. What took over was not the model prevalent before Thales' time in Miletus, but in Jerusalem.

In the Hebrew cosmogony every thing and every event is a direct consequence of the will of the anthropomorphic God. And since that will is not subject to any constraint, the Principle of Sufficient Reason does not apply. God is omnipotent—He can do absolutely anything. (There are logical problems with the conception of Omnipotence, but since, as we have noted, logic does not constrain divinity, they can be disregarded here.) Therefore, anything that is could have been otherwise, had God so chosen; and since His will is inscrutable, we can never *know* that things *must* persist as they are now or have been. This means that in terms of the distinctions introduced by Parmenides, *all* truths about the world are contingent, subject to revision, including Aristotelian "Natures."

This conception of what is possible contrasts starkly with that of the Greeks. They recognized only one kind of possibility, namely, what *could* be, whether or not it actually existed, *in the world as it is*. No round square or other self-contradictory entity was possible, but neither was Pegasus or the Chimaera.

In such thinking about the world as was still permitted to go on after Christianity became the state religion, the domain of necessary truths, the propositions describing states of affairs that "cannot be otherwise," shrank to those whose denials were self-contradictions. (Even further: St. Peter Damian, for example, held that God could make the past not to have been. But St. Thomas Aquinas's more moderate view prevailed.) Everything else became *"logically possible,"* as it was tendentiously christened; and imaginability, despite its subjective and variable nature, was held to be a sufficient condition for applying this designation. During the millennium when philosophy was "handmaid to theology" these notions of "logical possibility" and "law of nature" became so ingrained as to take their place in ordinary language, the unquestioned background of all thought, taken for granted, not noticed to be embodying any substantive doctrine at all, whether scientific or philosophical.

Law of Nature, implying the separability of what a thing *is* from what it *does*, together with Logical Possibility as a genuine kind of possibility, like physical possibility only broader, add up to the notion of *possible worlds*. This also was not envisaged by the ancients; nor could it have been. Some, such as Anaxagoras and Lucretius, thought that there were other inhabited lands out there in space; but that is not the same thing. Possible worlds, however—alternate worlds to ours which God in His omnipotence could have created had He seen fit—are perfectly congenial to the *Genesis* mode of thought.

As we shall see in later chapters, these two innovations in the manner of thinking about the world,

1. that anything imaginable can happen, should the Omnipotent Creator and Legislator (henceforth OCL) see fit to make it happen, therefore no informative statement about the world can be a necessary truth; and
2. that the regular behaviors of things are to be conceived and explained in terms of obedience to laws decreed by the OCL

would in the long run turn out to be the most persistent and crucially important modifications at the basis of philosophical thought that

Christianity imposed. But they were little noticed at the time if at all: perhaps because the first tenet, when made explicit, seems to concern only a technical logical point about the modality of scientific statements, and the second, "Law of Nature," was an easily understandable though rather rare metaphor, found indeed in Plato (*Timaeus* 83E), Aristotle (*On the Heavens* 268a13), Demosthenes (*Against Aristogeiton* I 65), and Plutarch (*Stoic Self-Contradictions* 1044C; *On Fate* 569e6). A more noticeable respect in which Christian doctrine diverged from pagan—*paganus,* cognate with "paysan," "peon," "peasant," meaning "country bumpkin," was the term that the Christians, who mostly were city folk, contemptuously applied to their opponents, who in their turn called the Christians "atheists"—conceptions was in regard to time. Unlike Greek notions of time as cyclical, for example, Plato's "Great Year," Empedocles' alternating Reigns of Love and Strife, time was held to be linear by St. Augustine, the principal philosopher (sort of) among the Fathers of the Church. It had a beginning: the Creation. It had a midpoint: the Incarnation of Christ and His suffering. And it would, after the Last Judgment, not exactly cease, but be succeeded by Eternity in which the Saved would rejoice, the Damned burn, and nothing else would ever happen.

In their ritual practices the Christians followed the Hebrew model in their abhorrence of idol worship. This was the point on which they got into the most serious trouble with the imperial government: burning a pinch of incense before the statue of a deified former emperor was to them idolatry, which many refused to perform even on pain of death. The emperors after long struggles had grudgingly excepted the Jews from the requirement, but did not extend it to the Christians.

But it was neither in doctrine nor in ritual that the most significant and ultimately decisive difference between the Christians and the rival mysteries was to be found. It was in their inheritance from the Jews of the Book: the Holy Bible. The world was full of religious texts, but none was venerated in the way that the Pentateuch was. It was held to be, literally, the Word of the One True Lord God Yahweh Himself, communicated by inspiration to scribes, long forgotten, who were nothing but His secretaries, and copied from age to age with the extremest care in rolls called Torahs. These five books were the Law, containing also the history of the world from the Creation to Moses. The same absolute authority pertained to the books of the Prophets that followed them in the text. There was nothing like this bibliolatry among the Greeks and Romans, or even other Near Eastern cultures. The Book became so to speak a portable homeland, supporting a coherent Jewish culture throughout the Roman world both before and after the destruction of the Temple of Jerusalem and consequent Diaspora.

Among these people, however, there were disputes and sects, mainly over the question of the immortality of the soul, about which the Law and the Prophets had nothing to say. The Pharisees were positive on this question, the Sadducees negative. The Christians took the Pharisaical position. St. Paul, the Jewish persecutor of Christians who after his famous epiphany on the road to Damascus became their principal advocate and in effect formulator of Christian doctrine, simplified the process of conversion by relaxing some of the more vexing requirements, notably circumcision. His tremendous abilities as an organizer created the Unified ("Catholic," i.e., "in relation to the whole") Church. His Epistles to various congregations, organizational but also explaining the significance of Jesus' crucifixion and resurrection, became the nucleus of what later, with the addition during the first century A.D. of the four Gospels, came to be called the New Testament, venerated even above the Law and the Prophets, the Old Testament.

THE TRIUMPH OF JERUSALEM

"What has Athens to do with Jerusalem? Or the Academy with the Church?" scornfully and rhetorically asked Tertullian, an apologist for Christianity in the time when it was still being persecuted. The Gospel of St. John, though indicating in its opening sentence—"In the beginning was the *Logos*, and the *Logos* was with God, and the *Logos* was God"—that its author had some acquaintance with Greek philosophy, probably Platonism, has no more to say about human reason, which St. Augustine when he came to consider it declared to be worthless unless pre-validated by God's grace. ("I believe in order to understand," as St. Anselm put it several centuries later; in other words, disagreement with Christian dogma was in itself proof of error.)

Explicit teachings in an infallible Book and the tight organization of the Church made possible something else unknown to the pagans: the notion of heresy as a crime. The word itself meant merely "opinion," but it came to be understood as opinion contrary to orthodoxy, "right belief." The sorting out of one from the other became a principal literary genre of the first centuries of the new era. (With the serendipitous consequence of preserving some pagan philosophy texts, since the disputants were much given to accusing one another of affinities to this or that pagan source, which they would quote) When orthodox Christianity became the state religion, heresy became a capital crime, for which in the succeeding thirteen centuries many thousands, perhaps millions, of persons were put to death by burning.

However, for a while the continued teaching of Greek philosophy was tolerated in the form of Neo-Platonism, a mystical doctrine with many affinities to Christianity (it and not the Holy Scriptures was the main source of the Christian conception of the soul and its immortality), in Alexandria until the year 431 and in Athens until 525. In 431, Hypatia, the head of the Neoplatonic school and the most eminent female philosopher, scientist, and mathematician of antiquity, was murdered with extreme brutality by a Christian mob instigated by St. Cyril, the local pope. In 525 the Emperor Justinian closed the Platonic Academy, which at the time, surprisingly, was flourishing under the leadership of the great scholar Simplicius, and forced the philosophers to seek refuge at the court of Persia, of all places. Justinian founded a sort of university in Constantinople, capital of the Eastern Empire, but in nearly a thousand years of existence it did not produce a single scientist or philosopher of note.

In the Western Empire, which never had been exactly a beehive of scientific or philosophical activity, it was easier to bring these to a halt. Indeed it could be said to have happened almost automatically, with the cessation of teaching of the Greek language, knowledge of which was deemed unneeded after St. Jerome had translated the New Testament into Latin. Nevertheless, Pope St. Gregory the Great went out of his way to help the process along by destroying the last public library in the city of Rome. Only in northern Italy was there some effort at preservation of the classics, by Cassiodorus, a wealthy Christian who founded a monastery on his property and set the monks to work copying books, and by Boethius, author of *The Consolations of Philosophy* and a high official at the court of the (German) Emperor Theodoric, who set out to translate the works of Aristotle into Latin. But he had completed only some of the logical writings when he was convicted of treason (a bad rap, it seems) and put to death.

In the extreme West, Ireland and England, some monasteries had continued to foster learning, so that at the beginning of the ninth century when Charlemagne, the first Holy Roman Emperor, wished to revive classical culture, teachers such as Alcuin, student of the Venerable Bede at the monastery of Jarrow in York, were available to help him. But the "Carolingian Renaissance" was not continued by his heir Louis the Pious; and anyway, the (still heathen) Norsemen descended upon Europe, creating anarchy virtually everywhere.

The seventh century had seen the sudden rise of the new religion of Islam, with its own Book, the Koran. The Muslims quickly conquered all of North Africa, which included a large part of the Eastern Roman Empire. In doing so they came into contact with the old Greek culture, the scientific (particularly medical and astronomical) and philosophical parts of which they found useful and absorbed, producing among themselves important scholars (of Aristotle, mostly) who, ironically, later passed on this revived knowledge to Christians in Spain. But after three centuries Muslim scholarship was strangled once and for all by imams, muftis, mullahs, and ayatollahs.

THE MEMBRANE REPLACED

The Christian clergy had to be educated in the Latin language and in theology. Schools for this purpose grew up around cathedral churches, some of which evolved into universities, typically having four "faculties," law, medicine, theology, and "arts," including philosophy, in which Aristotle could once again be studied. Some universities grew to great size: Paris is said to have had twenty thousand students.

Thus it was becoming known to more and more intelligent young people that there were attractive Grand Theories of Everything alternative to Christian orthodoxy. Pressures were building up that might have produced an intellectual explosion, had not the membrane separating high from low beliefs been reconstituted—this time visibly. St. Thomas Aquinas worked out a compromise according to which science—in effect, Aristotle—was to be recognized as primary authority in describing and explaining nature, which was everything, including the human body and its perceptive faculties, that Aristotle had treated of. But Aristotle, being a pagan, had been unaware of the realm of God and His angels, and of the human soul insofar as it is the immortal part of man with a destiny transcending this world. He had also been unaware of those truths, such as the creation out of nothing by the Word of God, that are knowable only by the revelations contained in the Holy Scriptures. So all this was to be left to the Church. This compromise staved off a major crunch for four centuries. Its philosophy, called Scholasticism, is still official Catholic doctrine and taught, more or less intact, in Catholic colleges and universities.

THE SECOND RUPTURE

But many Christian high beliefs are not very high, so to speak. The authority of the Church is based on texts which include some assertions capable of being put to checks in experience. When they give negative results, as sometimes happens, there ensues the "warfare of science with theology in Christendom," so brilliantly depicted in its strategies and tactics by Andrew Dickson White in his classic history of that name.

This war, which still continues, was far more serious and on an immensely larger scale than the conflict between Greek scientists and the gods of

Olympus in the days of Aristophanes. This time around, one side was the preserve and interest of a large, powerful, and organized priestly class, which had managed to make disagreement with its beliefs a capital crime. Science, on its side, was in some respects in a better position than the early Greek investigators of nature had been. First, there was not perceived to be any obvious conflict between investigation of "second causes" and orthodox religiosity; indeed, many eminent scientists were priests. Second, the scientific enterprise was on a better footing. By the seventeenth century not only had the corpus of Greek science been reclaimed, but in important respects it had been corrected and added to. Technological advances in the Middle Ages had produced precision instruments: clocks, telescopes, microscopes; calculating methods were revolutionized by the adoption of the "arabic" numerals; it was coming to be seen that scientific knowledge had practical applications, particularly in warfare, thus enlisting the support of the secular arm. And above all there was coming to be an appreciation of the role of experiment in the confirmation of scientific hypotheses—in "tethering."

Ironically, the decisive conflict came about over a high belief not introduced by Christianity, but simply a part of the universal naive view of the world: the fixity of the earth at the center of things. Naturally enough, the Holy Scriptures of the Jews presupposed it and referred to it here and there. By the sixteenth century, however, enough exact information about planetary motions had accumulated to lead Copernicus to hold that they could be more expeditiously explained by the double revolution of the earth. Counterintuitive as this hypothesis was, Galileo's telescopic observations of the moons of Jupiter and the phases of Venus tethered it further to low beliefs. The Church chose to take a stand on the issue. From our standpoint, it may seem that it was rash of the hierarchy to embroil themselves in this controversy. But if the earth is not at the center, and is insignificant in size compared to the rest of the cosmos, it is hard to continue to believe that the whole universe was created for the sake of one species of terrestrial mammal. Consequently, Galileo was perceived by some high Church officials as attacking the foundations of the moral order—as Socrates had been seen by Aristophanes.

The universe redescribed by Copernicus, Galileo, Kepler, and Newton could be reconciled with the picture found in the Holy Scriptures only by strenuously allegorical interpretation, supplemented by inattention. It was not only the physical pictures that were at variance; a deeper dissonance was the contrast between the frequent miraculous suspensions of natural laws, related in Scripture, and the view of nature, supported by science, as working in accordance with exceptionless regularities.

Since the Galileo affair, it has been downhill all the way. Some further blows to the basis of Christian high beliefs have been "higher criticism" of the Bible; geology; fossils; and Darwin. The ensemble makes it impossible to hold in the same mind both credence of the ascertained results of scientific investigations, and belief in the literal inerrancy of Scripture. Educated believers are therefore necessarily soft on the latter issue, having recourse to allegorical interpretations, excusing the language as targeting the imperfect understandings of early Israelites, or simply ignoring the whole issue as something for specialists to worry about; and sincerely professing themselves to be Christians while also believing what they hear from Big Bang cosmologists, paleontologists, and scholars of ancient Near Eastern literature.

Yet the question cannot be avoided, Why has Christianity not gone the way of phlogiston and phrenology?

Part of the answer is, of course, that it is more than a set of beliefs; it is an institution, an activity, for many still a way of life, with a tremendous grip on the emotions. Nobody had much of an emotional investment in phlogiston or phrenology, which in any case were hypotheses, not high beliefs with the prestige of tradition. But there is more:

If the basic contention of this book is correct, we are "hardwired" in two ways: first, to accept true low beliefs; second, to accept edifying high beliefs. When we lived "naturally," in small bands, occasions for these proclivities to come into conflict hardly ever arose, and when they did there were mechanisms for quieting the disturbances. But science—imaginative construction of general truths from low beliefs—produced strains that those mechanisms could not absorb.

We have described "believers" who manage to accept both science and Christianity. But of course there are many who, accepting science, reject traditional religion. If the contention about our nature as double believers is correct, we should expect that this condition would be unstable—that such persons would be attracted to other high belief systems not perceived to be vulnerable to undermining by science.

As everyone knows, this in fact happened. Many people embraced Marxism, which purported to be a science, but which was in reality a high belief system, indeed remarkably like Judeo-Christianity, one moreover that seemed capable of taking over traditional functions of religion such as validating an ethical code and holding out hope of salvation. It appeared to many that a truly scientific view of man and of his destiny was at last available.

However, it did not prove possible to inspire the masses with the requisite faith. Moreover, Marxist high beliefs, like Christianity's, were not high enough: they included all too definite predictions of what would happen in given circumstances, and the failures of things to work out according to the schema were too manifest. And so Marxism collapsed, less than a century after its first empowerment.

SUMMARY

In Greek culture, religious fervor attended not the official Olympian cults but the Mysteries, mostly importations from the East. They had initiations in which secrets were revealed, always including promises of immortality to the souls of believers.

A policy of religious toleration was general throughout the Hellenistic world, especially after its consolidation into the Roman Empire. But one religion was exceptional: Judaism, which held that there was truly only one God, creator, warlord, and legislator, Who accomplished everything simply by willing it—the ultimate development of the Command side of the human-action model of explanation.

Christianity was a Mystery derived from Judaism, retaining its beliefs about God, and Jewish intolerance of other religions. It became the Roman state religion and imposed this model on the intellectual world. Most important, the omnipotence attributed to God made everything about nature contingent; the domain of necessary truth now shrank to the triviality of mere denials of self-contradictory propositions. Everything else imaginable was now held to be "logically possible," and could be the case should God will it. And God having imposed the Laws of Nature and being able to suspend them—*miracles*—the distinction opened between what a thing *is* and what it *does*. Together these notions added up to the conception of "possible worlds," unknown to the ancients.

When Christianity acquired the power of the State, it forcibly suppressed most pagan teaching, which controlled no more institutions after 525. But gradually over the following six or seven centuries, Greek learning—aided, ironically, by the Muslim conquests—was revived to the point of being a rival to Christian orthodoxy.

St. Thomas Aquinas worked out a compromise: Aristotle was held authoritative as to the usual behavior of Nature, including the human body; but the human soul, and truths knowable only by revelation, were the province of the Church. Thus a new membrane was hung up between high and low beliefs, this time a visible one. But it was fairly soon ruptured, notably in the Galileo case.

Since then it has become difficult or impossible to take seriously both natural science and scriptural inerrancy. Christian beliefs, however, persist on account of the tremendous emotional investment in them. The Marxists tried but failed to establish a viable substitute.

CHAPTER 18

༄

Cartesianism

The breakdown of the so-called medieval synthesis, both political and intellectual, reached its climax (if that is the word) in the first half of the seventeenth century, the period of the Thirty Years' War between (mostly) Catholics and (mostly) Protestants fought (mainly) in Germany, the population of which it reduced by perhaps two-thirds. At every opportunity, Catholics, Lutherans, Calvinists, and Anabaptists, cheered on by Scholastics, Skeptics, Pietists, Fideists, Jesuits, Dominicans, Franciscans, Inquisitors, Realists, Nominalists, and Terminists, raped, hanged, starved, mutilated, beheaded, broke on the wheel, and burned one another alive, with the aid of racks, iron maidens, waterboards, the Boot, and other ingenious apparatus.

While this was going on, the smartest man in Europe, who only wanted some peace and quiet and had moved to Holland from Paris to get it, produced the first of the "modern" philosophies.

René Descartes (Latinized "Cartesius," hence Cartesian and Cartesianism) had been recognized as a genius since his boyhood at the Jesuit school of La Flèche. He was by far the foremost mathematician of the age, having early on invented the method of giving algebraic expression to geometric figures through their "Cartesian coordinates" when drawn on graph paper: analytic geometry. He had written the *Discourse on the Method*, a treatise on procedure in natural as well as mathematical science. And he had ready for publication *The World*, his Grand Theory of Everything, which, however, he prudently withheld from the press when he heard of the condemnation of Galileo.

For while Descartes accepted the main idea underlying the Aquinian compromise, that a human being is a soul in a body, he had a different interpretation of this dichotomy. An animal's body, he thought, is essentially a machine, as in the drawings of Leonardo da Vinci. It is made entirely from matter, Extended Substance; the muscles are like ropes, the joints like pulleys, and so on; its actions are in principle explainable entirely in mechanical terms. And so is the human body—up to a point. The human being possesses—or rather, *is*—a *mind* or *soul*, an immaterial entity created by God, a Thinking Substance ("The Ghost in the Machine," as Gilbert Ryle derisively called it). This entity is capable of altering the flow of the "animal spirits"—the fluid in the afferent nerves (Descartes of course thought in hydraulic, not electrical, terms) that bring information from the world to the brain, and in the efferent nerves initiate muscular motions. Having Free Will, the soul cannot be explained mechanically; and its actions are subject to moral appraisal, the business of the Church.

This Real Distinction of body and soul was of course basic Church dogma. But could it be proved? By a novel ploy—the use of Skepticism in the service of Faith—Descartes sought to do so in his *Meditations on First Philosophy*. It was not new, of course, to use Pyrrhonism to knock down rivals of Christianity. The apologists of the first centuries did it, and Pierre Daniel Huet, Bishop of Avranches, was doing it in Descartes's own time, but Descartes, in a sort of intellectual ju-jitsu, would employ skeptical premises in an endeavor to establish a positive, (allegedly) rational, and *certain* conclusion. He hoped that in gratitude for having done this, his teachers the Jesuits would abandon their suspicions of his orthodoxy.

Let us look again at the *Meditations on First Philosophy*—still in our day the inevitable textbook for Philosophy 101—in which he attempts to bring off his coup.

Descartes's strategy went Huet one better. The Pyrrhonists had doubted only high beliefs, whether tethered or not. They left alone the immediate deliverances of the senses ("it seems sweet"), the existence of a world distinct from our sensations and opinions, and mathematics—all low beliefs. Descartes purported to bring all beliefs whatsoever (save one) under suspicion. First, he can't be sure he isn't dreaming, for in dreams he has often been "deceived"; dream experiences, as experiences, can be indistinguishable from waking, and he can't be certain that now when he *seems* to be sitting by the fire and writing, he isn't in fact lying in bed. (Plato had made this point in the dialogue *Theaetetus*, but did not follow up on it.) But awake or asleep, he admits, two and three always add up to five, and a square always has four sides. To bring these beliefs within the domain of the doubtful, he has recourse to the famous suggestion that there might be "a malignant demon, who is at once exceedingly potent and deceitful, who has employed all his artifice to deceive me." This possibility, he avers, he can't rule out. So then, is any belief immune to doubt? Only that he is thinking (i.e., conscious—Descartes lumps believing, doubting, feeling, willing, etc., all together as "thinking"), consequently that he exists: "*Cogito, ergo sum.*" (This famous phrase is from the *Discourse on the Method,* not the *Meditations.*) Not even the Deceiving Demon could make it the case that he was mistaken about this, that he didn't *really* exist but only *thought* he did. But *only* that he exists as a thinker; he can still doubt whether he has arms and legs, and so on.

"Cartesian doubt" refers to this argument, presented in the first two *Meditations.* In the remaining four, by twice "proving" the existence of God and inferring that His perfect (by definition) benevolence would not allow anyone to be so horribly deceived, Descartes vindicates his (and therefore everybody's) knowledge of mathematics and natural science insofar as it consists of carefully considered "clear and distinct ideas." The real distinction of mind and body follows, he claims, from the fact that existence of the latter but not of the former is subject to doubt.

Why was—and is—this essay in Christian apologetics regarded as the first major document in "modern" philosophy? Well, it began the New Way of Ideas, of which we shall have more to say . Further, in his "real distinction" of mind from body, Descartes did not draw the line where either the Greeks or the medievals had; whereas they had put sense perception on the body side of the line, Descartes's mind or soul includes them, for they are part of consciousness, subjectivity. (Dogs and cats not having souls, they therefore lacked consciousness too; they were simply stimulus/response mechanisms, and it was OK to vivisect them. They felt no pain despite their howls and screams, which were compared to the squeaks of unlubricated machinery.)

But probably the main reason for awarding the primacy of modernity to Descartes, and for the Church's continued hostility—in 1664 all the works of Descartes were placed on the Index of Forbidden Books—lies in the skeptical cast that his strategy imparted to philosophizing. A near consensus of "modern" philosophers believe that Descartes's negative moves were more convincing than the positive; and they continue to be scared by the Evil Demon. In consequence, philosophy since Descartes's time has been obsessed with the alleged problem he left behind, of how to get from subjectivity to objective knowledge, as well as a host of other "problems of philosophy" unknown to the ancients but flowing from this one such as The External World, Other Minds, and Personal Identity.

All these considerations arise from Descartes's use of the Evil Demon ploy.

Where did that unpleasant character come from, and did Descartes *have* to bring him in to cast suspicion on his belief that $2 + 3 = 5$?

It is hard to see how he could not have. The ancients, including the Pyrrhonists, recognized mathematics, at least of the simpler sort, to consist of necessary truths, propositions that "could not be otherwise"; one who did not believe them would have to be ignorant or insane. The reason for this was that they were the lowest of low beliefs, rubbed up against reality in every encounter in which they figured, and indispensable for coping. In the absence of familiarity with the notion of an omnipotent being, the suggestion that such beliefs could be mistaken would be entirely gratuitous, could have no motivation, and if it were ever put forward would be dismissed out of hand. It would be like suggesting, out of the blue, that everybody, including oneself, is crazy.

True, an agnostic or atheist at the present day reading the second *Meditation* may find the argument interesting and even compelling; but that is because despite his or her beliefs, this notion of the omnipotent being is now virtually inexpugnable as part of the background of thought in (however former) Christendom.

The Demon scenario also depends on medieval theology in another way, assuming as it does that the omnipotent being can get into a mortal's mind and implant therein what thoughts he or she pleases. In the *Iliad*, to be sure, Zeus sends false dreams; but that is not relevant to what is at issue here. Dreams have already been admitted to be "deceptive"; the Demon has to be brought in to cast doubt on those ideas that Descartes so far cannot doubt *whether awake or asleep*. In the philosophies of the Cartesians Berkeley and Leibniz, however, causation by divine fiat—including the contents of minds—becomes the sole modus operandi in the universe.

This is our affirmative answer to the historical question of whether Cartesian skepticism depends on medieval theology. Let us now consider the logical question of whether Descartes's scenario of the evil demon, regardless of how it was generated, bequeathed a genuine problem to philosophy.

It did not, despite the virtually unanimous consensus of philosophers that it did. The Evil Demon ploy demands that before we can justifiably claim to know that *p*, we must already, before making that claim, have ruled out as impossible every circumstance such that if it obtained, *p* would be false; and, it avers, Here is one such circumstance. But we *can*, antecedently to any inquiry, rule out the possibility of our being deceived by an evil demon. This is because we know that there are no evil demons. {See Leite 2010.} No question is begged in this response, even though if we *were* being deceived by an evil demon, we wouldn't know it, by hypothesis. For as things stand, we have overwhelming evidence that there are no evil demons, and none at all that there are or could be. Unless our worldview *already* allowed a priori for the existence of omnipotent beings, there could be no reason whatsoever for raising this question.

Another consequence of Cartesian skepticism, and one not cured by the author's affirmation of the external world in the happy ending of the *Meditations*, was the substitution of "ideas" for things as the objects of perception. Since, Descartes argued, we can have a visual experience indistinguishable from that of looking at a tree, even though there is no tree there, what we really, directly, and immediately see when we think we see a tree is not after all a tree, but the "idea" of a tree (later called a "sense datum"), which may or may not be like a tree out there in this or that respect; and it becomes a philosophical problem to determine what these respects are. The theory of perception is thus moved from optics into philosophy; and once this is done, the bold view of Bishop Berkeley—that everything *is* a show being put on in our minds by God, Who is just like the Evil Demon only nicer—seems inescapable in the end. But there is no need to go along with Descartes here either. {See Austin 1961, 1962a and b; Matson 1976, chapter 2.} Very briefly, the reason for this is that whatever may be going on in the brain (or mind), it is not a further instance of something being seen; it is the seeing itself. In our terminology, the internal indicator is not a picture, whether accurate or distorted, of what it indicates, much less the spectator of such a picture.

Henceforth in this book I shall call "Cartesian" any Grand Theory that incorporates the contingency-of-the-world doctrine, and "Milesian" any that does not. The textbook classification of "Continental Rationalists" versus "British Empiricists" is ludicrously inept. It ignores important affinities where they exist, and suggests them where they don't. And it leaves out Hobbes, a major figure. {See Matson 2000, 435–436.}

LEIBNIZ

Leibniz's doctrine of Monads, being basically derived from Plato's teaching that only souls are simple beings, cannot be blamed entirely on medieval theology. But the Pre-Established Harmony can. It requires not merely an omnipotent being but one who is also omniscient and benevolent.

Leibniz "proves" the existence of a God having these properties. His proof, however, is based on the contingency of the world—for which he offers no argument, taking it as a self-evident axiom. But this is exactly what is at issue. So the argument is circular, as well as derived from medieval theology.

Leibniz's entire metaphysics collapses into ruin if the Omnipotent and Omniscient God is pulled out from under it. In particular, "possible worlds," as Leibniz conceived of them—God's thoughts, which must be real entities if there are any such things—vanish. With them also disappears "logical possibility," if conceived as anything more robust than mere absence of contradiction.

How are we to tell whether a notion is free of contradiction? In simple cases, this is not a problem. A purported state of affairs is self-contradictory if among the propositions needed to describe it, one is the denial of another: for example, "It is day and it is night." A concept is self-contradictory if its definition contains logically incompatible elements: for example, "An apatagon is a plane figure having exactly three straight sides and exactly four interior angles."

But what about more complicated cases? The best we can do, so we are told if we press the matter, is to try to imagine the concept or state of affairs. If we succeed in imagining it, then it is OK.

This criterion is another inheritance from medieval theology.

On the face of it, to make the ability to imagine something—a psychological and subjective accomplishment, at which some people are more skillful than others—the test of a logical status cannot be satisfactory. Some people say they can imagine time travel, the non-existence of God, locomotives moving without horses inside; others say they cannot. And it is doubtful whether anyone can imagine the Big Bang, twelve-dimensional strings, or the number aleph-sub-zero, but that does not keep these concepts from playing their roles in respectable theories.

For these reasons, any argument with a premise of the form "X is imaginable" and concluding that a certain state of affairs obtains, out there in the real world, ought to be rejected, or at least to come under strong suspicion. This is the pattern of the famous Ontological Argument for the existence of God of St. Anselm and of Descartes (in the fifth *Meditation*), which besides is yet another consequence of the Christian conception of the Omnipotent Being. That sophism has been sufficiently exposed, by Kant, Hume, and others. But they have only scotched the snake, not killed it, for its mode of argumentation is still around—for example, in David Chalmers's mind-body dualism, based entirely on the author's claim that he can imagine a zombie. (Therefore zombies are "logically possible," therefore there are "possible worlds" containing zombies, therefore—this is the "ontological" move, from mere imaginative conception to a feature of the actual world—consciousness *is* separable from, not supervenient upon, physical structure and behavior.) On this basis, Chalmers advocates that scientific psychology should turn its attention to the study of the merely contingent "bridge laws" correlating consciousness with physiology. Yet so astute a critic as John Searle can only make fun of the theory, not refute it, because he too admits the "logical possibility" of the zombie. {See Chalmers 1996; Searle and Chalmers.}

SUMMARY

Cartesian skepticism, unlike Pyrrhonism, was total, calling into question low beliefs as well as high. Descartes himself was not a skeptic but set out the argument in its favor for the purpose of refuting it and thereby strengthening theology. His argument was only possible against a specifically medieval background, his Evil Demon being the OCL in disguise. But as the skepticism was more convincing than the refutation, this concept is still around in our day, responsible for "modern" philosophy's obsession with finding "foundations" for knowledge. So is the concept of causation by direct divine fiat. Greek gods didn't operate that way.

Leibniz's Pre-established Harmony also depends on the OCL. His philosophy collapses if you subtract this Being from it—"possible worlds," "logical possibility," and all.

Imaginability, making subjective conviction the measure of logical status, is unsatisfactory as the criterion of possibility.

The Ontological Argument for the existence of God (Anselm, Descartes), which likewise moves from subjective conceivability to objective existence, is nowadays usually considered to have been shown up as fallacious. But its pattern can still be discerned in David Chalmers's advocacy of mind-body dualism: the subjective "logical possibility" of zombies purporting to show the objective reality of the schism.

CHAPTER 19

✿

Miletus Preserved I: Hobbes

It is evident that science, after coming to maturity in ancient Greece, has always proceeded in accord with the three Milesian requirements of Unity, Immanence, and Reason. It has not *presupposed* them—science has no presuppositions—but they are its framework. One would naturally suppose, then, that "modern" philosophical Grand Theories would likewise put these among their central tenets, or take them for granted as their starting points. Yet such has not always been the case, as we have just seen in considering Descartes and Leibniz. Happily, however, two seventeenth-century philosophers upheld the Milesian tradition through a period of general hostility to it: the Englishman Thomas Hobbes and the Dutch Jew Benedict Spinoza.

Thomas Hobbes of Malmesbury was the true founder of modern philosophy adequate to modern science. Along with Spinoza, he was virtually the immediate successor of Aristotle, as Galileo was of Archimedes—both consciously and unconsciously rejecting the entire "medieval interlude," in particular, the doctrine of the contingency of the world. Hobbes's frequent denunciations of Aristotle are (except for those aimed at the doctrine of Final Causes) really directed against the Supreme Infallible Authority that Aquinas and the Scholastics had tried to turn him into. Again and again Hobbes explicitly asserted that the truths ascertained by science are necessary, cannot be otherwise. This break with the past was as sharp and as absolute for his time and place as Thales' was with his. And it too began with a mathematical epiphany. We have a description of it from his biographer John Aubrey:

Being in a Gentleman's Library, Euclid's Elements lay open, and 'twas the [47th Proposition of Book I—the Pythagorean Theorem]. He read the Proposition. *By G——,* sayd he (he would now and then sweare an emphaticall Oath by way of emphasis) *this is impossible!* So he reads the Demonstration of it, which referred him back to such a Proposition, which proposition he read. That referred him back to another, which he also read. And so on that at last he was demonstratively convinced of that trueth. This made him in love with Geometry.

About the same time there occurred a conversation about the senses, in which someone asked "But what is sensation?" and no one present could reply. The answer Hobbes at length gave—that sensation is motion in the brain—is the basis of his conception of human nature.

Hobbes was popularly reviled as an atheist, and after the Great Fire of London, which the religious interpreted as divine punishment for the sins of the people, "in Parliament... some of the Bishops made a Motion to have the good old Gentleman burn't for a Heretique." King Charles II, who in exile had had Hobbes as his tutor in geometry, protected him, but he was forbidden to publish any more in England.

Hobbes outlines clearly the procedure to be followed in establishing necessary truths about the world. He makes no distinction between philosophy and "ratiocination," his word for science. Both consist in finding out causes from effects, and vice versa. And what there is to ratiocinate about is starkly simple: bodies in motion; that's all there is. In principle this includes the motions in the brain, which are the thoughts and "passions" (we now say emotions) at the base of ethics and politics; but in practice they must be studied by other means: observation and introspection—for human beings are roughly equal and alike in their mental powers, so that what I find out about myself can be extrapolated to others.

These remarks bear primarily on Hobbes's materialist Grand Theory as expounded in *De Corpore (On Body),* 1642, written in Latin and addressed to the learned. His popular reputation, however, was based on *Leviathan* (1651), which after a brief summary of these principles passes on to his political theory. This is, in our terms, an attempt to base a defense of absolute government on low beliefs exclusively, ignoring an alleged divine right of kings and other high beliefs usually appealed to in such endeavors. (Consequently, the book was publicly burned at Oxford after the Restoration.)

Although by instilling scientific interests in the young King in exile Hobbes played an important part in causing the Royal Society to come into being, he never was elected a member, because of an unseemly quarrel over circle-squaring. He enjoyed a peaceful and still productive old age as an honored retainer of the powerful Cavendish family, dying in his ninety-second year.

SUMMARY

Although participants in the scientific enterprise, today as in the past, comply at least tacitly with the Milesian requirements of monism, naturalism, and rationalism, not all philosophers have followed suit. Many eminent metaphysicians continue to believe in the contingency of the world, thereby still stirring the pot in which simmer the traditional "problems" of external-world skepticism, induction, other minds, and so on. But two seventeenth-century philosophers, Thomas Hobbes of Malmesbury and Benedict Spinoza, produced Grand Theories fully embodying the three Milesian requirements, and were persecuted for doing so.

Hobbes's Monism consisted in asserting that nothing exists but bodies in motion: reductive materialism. Sensations and all mental items are motions in the brain and ideally should be studied as such. He did not deny the existence of God, at least not openly, but was forced to admit that He too had to be a body. Hobbesian Naturalism was simply the claim that all bodies are and always have been in motion, not driven by forces other than those inherent in them. He was a Rationalist, holding that all true propositions are necessarily true, their necessity to be shown by the hypothetico-deductive method, of which he gave an accurate sketch. To him there was no distinction between philosophy and science: both are the finding out of causes from effects and vice versa, by "ratiocination."

Hobbes's popular reputation, however, was and is mostly as a political theorist, who attempted to justify an absolute sovereignty not by high beliefs such as Divine Right of Kings but from low beliefs about the necessary conditions for social living. It seems appropriate, though digressive, to devote the following chapter to consideration of what bearing the theory of high and low beliefs may have on our views about social and political interactions.

CHAPTER 20

∾

Institutions

I Authorise and give up my Right of Governing my self, to this Man, or
to this Assembly of men, on this condition, that thou give up thy Right
to him, and Authorise all his Actions in like manner. This done, the
Multitude so united in one Person, is called a COMMON-WEALTH.... This
is the Generation of that great LEVIATHAN, or rather (to speake more
reverently) of that Mortall God, to which wee owe under the Immortal
God, our peace and defence. (Leviathan Chapter 17)

In modern terminology, Hobbes is here describing the speech-act
creating an institution, the State. Since Hobbes is the first modern philosopher to attempt a theory based only on low beliefs, it is appropriate to
pause here and look into how the special kind of belief involved in these
matters—namely, what we may call second-order beliefs, beliefs about
beliefs—works.

Some beasts live in societies, with hierarchies (pecking orders, e.g.),
duties, ritualized combats, and so on; but a walrus's harem is not exactly the
same sort of organization as a sultan's. The main difference is that the latter
is set up by *constitutive rules* expressing a *collective intention* of the persons
involved as to who shall belong to the harem, what kinds of conduct are
allowable in it and what forbidden, and so on. It then becomes, for instance,
a fact—described in a true proposition—that it is a *crime* for any male
person other than the Sultan or a eunuch to enter the enclosure; but an
institutional fact, depending for its existence on agreements and beliefs.

What amounted to institutions and institutional facts (henceforth
"i-facts") were a major topic of discussion in Greece as early as the fifth
century B.C. This was the controversy over what was "by nature" and what

"by convention." The most famous of these was the first book of Plato's *Republic*, in which the sophist Thrasymachus is depicted as maintaining that Justice or Righteousness is merely a name that the power elite of a society have bestowed on behavior that is in *their* interest (as opposed to the interests of those lower down the totem pole). They make laws to this end, calling them "just," and get the foolish common people to observe them. (For this theory Thrasymachus deserves to be called the first "postmodernist." It is a recurrent view, whose chief advocate in the twentieth century was Michel Foucault.) Plato himself proposed the classification of inhabitants of his ideal city into three castes on the basis of a "noble lie": the Myth of Metals according to which those with gold in their souls should be guardians, silver soldiers, and bronze workers. These institutions would create i-facts, for with them in place it would be a *fact* that someone violating the law would be acting unjustly; or that soldiers and workers owed deference to guardians.

The Searlean formula for the speech-act that creates an institution is "X counts as Y in context C." This is a *constitutive rule.* For instance, "This piece of wood carved in the shape of a horse's head counts as a knight in the game of chess." "This piece of paper with a portrait of Abraham Lincoln on it counts as $5 in the context of buying and selling." It then can be an i-fact that you owe me $5. You can't play chess without agreeing to the first rule, nor get very far in commerce without accepting the second. Your agreements, however, are unlikely to be conscious—this is just the way things are; the distinction between natural facts and i-facts ordinarily goes unnoticed, just like that between high and low beliefs.

A moment's reflection will convince you that i-facts are pervasive in human society—indeed, language itself is the first institution, and every fact about it is an i-fact. The ability to acquire a language is not an i-fact, but which noises bear what meanings is. Nevertheless, given rules of syntax and semantics, observance of them becomes a factual requirement for meaningful speech, as we saw in our example of Lucy communicating the position of the aurochs. In one way, the difference between natural facts and i-facts is that the former are independent of beliefs, whereas the latter in general are not. What makes chess possible is not, strictly, *belief* that the horse-head is a knight, but the *resolution* that it is to count as such. And a resolution is not a belief, though *that a resolution has been made* may be a belief. To play chess, it is not enough for me to make the knight resolution; my opponent must make it too, and on top of that we each must believe that the other has.

In the case of the $5 bill, though, and of language, it makes no sense to speak of me, individually, making a resolution, and then waiting around for others to resolve likewise. For trade and talk to work, I must believe that everybody else *already* has the same belief that I have, however they may have acquired it. Thus these institutions depend on beliefs about beliefs already there—*standing* beliefs. The only reality that my belief that the piece of paper counts as $5 can rub up against is such a standing belief, unanimous or nearly so.

A CLASSIFICATION OF MEANS WHEREBY INSTITUTIONS ARE CREATED

1. Explicit resolutions. Examples: Rules of games. Who can make them: Anyone who wants to play the game; in case of "standard" games (chess, bridge, golf, ...) usually an Association, formed in various ways and with membership determined by another set of rules. Context: the game (actually being played).
2. Tradition, a vague term signifying unknown historical causes. Examples: marriage, property, various hierarchies. Particularly apt to be confused with natural facts and purported to be effects of supernatural causes.
3. Legislation.
4. Rights. Examples: life, liberty, pursuit of happiness. These come close to natural facts, because differential and softer treatment of members of a gregarious band is often instinctive. But in human societies, all rights are institutional. Who can make them? Anyone strong, influential, persistent, and eloquent enough to agitate successfully for them. As with traditions, recognition of the fact that they are institutional may be detrimental to their effectiveness—as recognition of a high belief *as* high may deflate it. "Human rights," supposedly prior to the organized State, is cant.
5. The Economy. All business dealings except barter (which, however includes transactions with currencies backed by gold or cowrie shells or whatnot) are in the realm of i-facts. The science of Economics, however, like all sciences, is not. Examples: credit, investment, property, fiat money, contracts. Who can make them? Dealers, and others, empowered by legislators.

ANOTHER INVISIBLE MEMBRANE

The institutional nature of rules of games, parking laws, the interest rate, and the right to bring up one's children as Catholic or Holy Roller, is recognized by everyone. However, the urge to imbed favored institutions in the very nature of things is powerful. Consequently, myths of origin are rife: they couldn't have "just growed"; there must have been a supernatural Institutor. This is especially so in cultures influenced by the Holy Bible, which made God the supreme legislator. Marriage (though of unspecified numbers), priesthood, the Sabbath, allowable and forbidden foods, circumcision, punishments, sacrifices, and so on are therein prescribed, thus blurring if not abolishing the distinction between nature and institution. Indeed, this theological account amounts to reconceiving nature *as* an institution, at least for the chosen people.

In modern "secular" societies, some of this attitude has carried over, like the contingency of the world. For instance, at the time of this writing the nation is distracted by the controversy over "gay marriage." Since many who are opposed to the notion nevertheless do not object to the granting of fully equal spousal rights to gay couples, it seems to come down, for them at least, to one point only, namely, whether the *word* "marriage" is to be allowed to occur in the speech-act creating a certain kind of partnership between persons of the same sex. Yet the emotions aroused are of tremendous intensity.

HOW ONE "JOINS" AN INSTITUTION

Another dimension into which institutions may be divided is compulsory/optional. Rules of games are examples of the latter. Where the playing of games is voluntary, you need not have any concern with their rules if you do not care to play. Similarly with the institutions of fashion, though here there may be social pressure to conform, in various degrees and relative to how weird one appears if not conforming. And how the rules of fashion come to be what they are raises interesting questions too; for obviously they are not created by speech-acts, nor could they be, except in the case of official uniforms and the like. There is no "collective intention" to make skirts lower this year. "Follow-the-leader" seems to be the usual origin—do as Yves St. Laurent or Coco Chanel, or whoever, does. So there is at least one means of generating an institution besides the Searlean speech-act. Does it involve belief? Yes. It can only happen when there is consensus as to who *is* the leader.

Compulsory institutions are those in which one holds membership like it or not. Language; the law (note that every crime, even when *malum in se*, is an institutional fact); money; marriage; property; citizenship; . . . These are the institutions most likely to be behind the invisible membrane, that is, generally thought to be "natural," not institutional at all.

Then there are matters about which there may be genuine controversy as to whether they are institutional or natural. The most important of these concerns the notion of Rights.

RIGHTS

There are no such things as natural rights. All rights are institutional.

This is not to deny the existence of the "right of nature," as Hobbes calls the "right" to self-defense. But that is a "right" without a correlative duty, so is not relevant here, where by a right is meant a restriction *on other people* of the license to do with and to me whatever they will, or a duty on their part to aid me in some range of my endeavors. Examples: life, liberty, the pursuit of happiness, property, inheritance, freedom of speech, education, equality, a living, three weeks' paid vacation per year. Jefferson said we were "endowed by the Creator" with the first three, which was effective rhetoric but dubious philosophy. He said that governments exist to defend these rights, and indeed a government that is indifferent to or obstructive of them is vulnerable to criticism; but that does not annul the fact that it is government (the institution of the power elite) that creates them, where they exist. A woman does not, in fact, have a right to drive a car in Saudi Arabia. Tough.

Where does rights talk come from? There is no word in Greek for "right" in the above sense. There are "the things that are one's own," *ta heautou*, the "due" in the formula for justice, "rendering to each his due," and perhaps this implies the notion. Yet there were opponents of the institution of slavery in Greece, who, however, said nothing about a right to liberty, only that slavery was against nature, therefore unjust, because the slave was the natural equal of the master. So Aristotle defended slavery by denying this claim, which fact is evidence for his not having had a concept of a "natural right," for this defense would not go through were such a concept involved. In Latin also there is no such special word; the closest one finds is, parallel to Greek, *jus*, "just." "I have a right to liberty" would have to be rendered as *"jus est me liberum esse,"* "it is just that I be free," which is not the same thing, for a natural right overrides even justice.

It appears that the first usages of "right" in the modern sense were during the Middle Ages, notably in Magna Charta, in connection with limitations to be placed on personages in the complicated relationships engendered by the feudal system: your liege lord could do this, but not that; you had a "right" to his forbearance in this respect. In Magna Charta this sort of limitation was extended to all freemen in the right to the writ of habeas corpus and to "due process of law."

If this account is correct, the concept of a right is not, after all, of theological derivation but arises from politics. What the concept is now, however, beyond specific restraints explicitly stated in valid legislation, is puzzling.

POLITICAL OBLIGATION

From the natural facts of being born in a certain territory and continuing to reside there, arise such i-facts as being obliged to pay certain taxes and to serve in the armed forces, and empowerment to vote in elections of public officials. This is because the place in question is not mere acreage but part of the physical base of an institution, the State. How does this situation come about? Purported answers to this question are theories of political obligation.

In some cases, for instance in the United States of 2011, a Searle type explanation may seem adequate: Certain persons in Philadelphia performed certain speech-acts setting up this institution. They were the appropriate persons to do this because of previous speech-acts declaring that they were to count as representatives to a convention to revise the Articles of Confederation. But already it begins to look as if we are embarking on an endless regress.

The favored regress stoppers have been theological—"The powers that be are ordained of God"—and Consent theories, to the effect that *all* affected persons have consented, at any rate *virtually*: they *would* (have) consent(ed) if they knew all the (natural) facts and, being rational, acted to further their own interests (Hobbes, Locke, Rousseau, Rawls).

Now, the only alternative to consent theories seems to be the establishment of the State by brute force, legitimized by nothing but the passage of time. ("Just growing.") And looked at historically, it is evident that many, perhaps all, states have originated this way, making them in their ultimate beginnings merely the natural extensions of pecking orders. Yet it is also clear that at least some present-day states can, on the whole, claim to enjoy the consent of the governed, even if it is only a sort of weary resignation—namely, those in which there is no revolutionary movement powerful enough to be a cause of concern. And whatever the facts, it is interesting to ask whether political institutions *could* originate and be justified by unforced consent.

The classical consent theories, those of Locke, Hobbes, Rousseau, and Rawls, appear to be inadequate. Hobbes's contractors set up so monstrous and arbitrary a Sovereign—not even a party to the contract—that this notably secular philosopher has to postulate a quasi-religious experience, "more than consent" {see Matson}, to motivate obedience to him. Rousseau attempts not a historical speculation but a specification of what would have to be the case if the State could be *justified*, and to accomplish this he has to rely on a rather metaphysical distinction between "general will" and "will of all." His notion that the citizen must be "forced to be free" seems to preclude even the possibility of there being such things as rights. Worst of all is Rawls's scary professorial fantasy of utterly inhuman legislators behind "veil of ignorance" conditions insouciantly devised to "get the desired result" for the institutions of social justice.

Nevertheless, contract theory is not hopeless. Although Locke's will not do as it stands, on account of the *logical* defect of starting out *in medias res,* that is, of assuming that the principal institutions that the State exists to defend— property and rights—are already there in advance of the State. It is like explaining the origin of baseball in terms of creating a game to give effectiveness to innings, runs, strikes, and balls. But this shortcoming can be fixed.

The following is a sketch of a revised Lockean scenario in which actual human beings, even if unequal in power, talent, and virtue, might nevertheless set up a viable State by consent.

HOW TO MAKE A STATE

To keep things simple, let us consider two persons only, named Robbie and Friday, on one otherwise uninhabited island. With salvage from the shipwreck that deposited him there, various skills, and hard work, Robbie has made himself a comfortable home. Friday, of another race but able to speak Robbie's language, arrived later with no possessions, in a canoe blown off course. In another telling of this story, Friday cheerfully subordinates himself completely to Robbie and they live happily together until rescued. Let us be more realistic.

Robbie's problem is to avoid being killed by Friday. Friday's is to get enough food, drink, and other necessities from Robbie—there is no other source—to survive. What are they to do? Well, the most obvious action would be for them to fight it out. But as both are without weapons, and of roughly equal stature and musculature, the outcome of that would be risky. So they sit down, some distance apart, and talk things over.

Friday: "You have plenty of food in your fridge. Just let me have half of it, and I will let you alone."

Robbie: "But I couldn't get by on just half. And neither could you, after a day or two."

At this stage, if they are not to fight, they recognize that there must be more food production. Friday: "I still have my canoe" (Robbie has no boat) "and I'm good at fishing. Give me a sandwich to tide me over, and I'll go out and catch us a mess of humuhumunukunukuapuaa." Robbie: "OK."

After their supper, negotiations might go several different ways. They might become friends and share and share alike. But that would be dangerous for both, as depending for its effectiveness on notoriously unstable emotions. Let us consider the worst case scenario: they already dislike each other intensely. Yet they realize that each would be better off if they could come to some arrangement assuring continuing peace between them. So they declare a truce while they try to devise one.

Robbie is in the better negotiating position. He already has his house, his workshop, and his farm. He will set to work—maybe even, with more sandwiches, getting Friday to help him—erecting a stout wall around the most essential area of his holdings, which he will then festoon with ingenious warning devices against trespassing. Unfortunately for him, however, one of his farm plots is too distant from the others to protect in this way; and it is where he grows veggies without which he cannot survive. Moreover, he has developed a taste for grilled humuhumunukunukuapuaa. So Friday has some bargaining power after all.

They can then agree to a non-aggression pact. "I, Robbie, will not attack you, Friday, provided that you reciprocate, and moreover do not come within my fences without my permission, nor take away items outside them that I have tagged. We will exchange fish for veggies at the rate of one pound for two. Additions and changes to this pact may be made by mutual agreement." This done, they formally shake hands and retire to their separate bunks.

A good day's work. They have created property (the stuff inside Robbie's stockade), rights (not to be attacked, to negotiate further agreements), an economy (exchange of foodstuffs), and the State (the association thus defined). Moreover, they have gone from Is to Ought in the Searlean manner: *pacta sunt servanda*; each party to the agreement *ought* to observe it. Is this "ought" merely prudential? Well, yes, but so are all oughts in the end.

Whoa, wait a minute, you may say. Did not a famous philosopher observe that "covenants without the sword are but words, and of no power to compel anyone"? True. But the parties *do* have their swords—their muscles, plus the pikes, spears, bows and arrows, and so on, that they will immediately set to work making. Undoubtedly, this pact—or truce, if you prefer to call it that—will last only as long as it is in the considered best interest of both parties to observe it. But that is enough. Or if it isn't, then War—the avoidance of which, remember, is the whole point of the proceedings—will ensue. But that is the human predicament.

Should other castaways show up, Robbie and Friday may be expected to try (probably with success) to get them to adhere to the pact, perhaps with revisions. And to avoid such "inconveniences" as Locke noted where there is no regular machinery for enforcement, police will be appointed. But that ceremony should not be regarded as the origin of rights and the State. The original compact should be.

Finally, it might be observed that the agreement would not necessarily have to be explicitly formulated in words; it might have been tacit, as long as it was really understood by the people involved. In that way the State could have "just growed."

BELIEFS ABOUT I-FACTS

Now back to the question about beliefs about i-facts.

I-facts are, after all, facts, so as such they can be part or even the whole of the reality that a hypothesis or belief has to rub up against to become a low belief. And there are false i-beliefs: errors about the law, for example, and all the beliefs of theologians. So it looks as if our general characterization of belief as an internal indicator assented to will apply without modification to i-beliefs, will it not?

Yet this can't be the whole truth; there is something special about i-beliefs just *qua* beliefs. Bolero jackets cease to be fashionable if people no longer believe that they are; you have a right to be hanged with a silken cord if your brother the Duke predeceases you. Does all this follow from the (natural) fact that all *values* are i-facts? And *are* they?

No, that can't be right. Values precede institutions; jellyfish have values. Life *simpliciter* creates them.

Another tack: Is it the *collective intentionality*, explicit or tacit, in i-facts that is the special ingredient? This looks more promising. Its disappearance accounts for the first case (fashion), its division for the second, and the third is perhaps too trivial to count.

A non-trivial further example: the Economy. The natural facts underlying the economy are things like weather, rain, availability of labor, fertility of the soil, transportation. None of these changed globally and disastrously in the past few years. Yet the Economy lost perhaps half its value (or price, if that is different). How? On account of i-facts, which as we have noted are all the economic facts there are past barter. The economic collapse is said vaguely to have been due to "lack of confidence." This is of course a matter of belief. Is there at work something we might call the Python effect, after the Monty Python TV episode about the high-rise apartment building that began to lean whenever its tenants doubted its structural integrity? How would this work?

Credit, from Latin *credo*, "I believe." The basic credit transaction is this: I lend you this amount of money *believing that* at a date certain you will pay it back with interest. This belief may be a matter of degree; that is the risk factor. I estimate it, ordinarily, on the basis of your reputation, that is, the summation of beliefs about you of people who have had dealings with you and know about your assets and liabilities. That is the reality that my belief in your creditworthiness has to rub up against. And my success as a businessman is thus bound up with yours.

But the system changed fundamentally in the last century. It became common for loans to be "bundled" and sold to third parties, who had no direct knowledge of the reputations (beliefs) involved, only statistical estimates. This fact, coupled with a strong tendency to relax requirements for creditworthiness, led to many of the bundles becoming "toxic assets," uncashable at their purported value because of high default rates among the borrowers. Unable to pay them off on demand, the holders were forced into bankruptcy, causing sharp revisions of standards among lenders so that credit dried up, with the consequences to normal buying and selling that we are now experiencing. This is indeed a case illustrating the Python effect: the banking system no longer depending just on beliefs about the creditworthiness of individuals, but of beliefs about these beliefs, which turned out to be false when rubbed up against the reality of cashing in.

THE REAL PROBLEM OF INDUCTION

A major argument of this book contends that the so-called Problem of Induction—how reason can justify inferences from the experienced to the inexperienced—is bogus. Nevertheless, there is a real and (I fear) insoluble problem of induction deserving that name, though it is about people and their beliefs, not nature as a whole; and practical, not philosophical. In our original setting up of what a belief is and how it may get rubbed up against reality, we used as our paradigm object of belief the animal's sizing up of "how things are going"—the dynamic situation of the hunter, the fighter, the object of pursuit, the would-be mate. But sizing up involves the element of anticipation, therefore of the future. We need now in our analysis to note this fact and take it into account when discussing the beliefs of persons. For where people and their beliefs, therefore their choices, are concerned, the future does *not* always resemble the past. From the fact that A has believed p up to t_n it does not follow that he or she will believe p at $t_{(n+m)}$). People are wary of this in respect to such institutions as fashions and politics, but it tends to get overlooked when it comes to matters of mass taking-for-granteds (which are beliefs) such as in the soundness of the banking system. Hence crashes, bursting bubbles, Python effects in general. These are natural consequences also of the propensity to believe what everybody else believes (or is thought to believe). Indeed, a stock market bears an uncanny resemblance to Plato's Cave, where the prisoners awarded prizes to those who made the best "predictions" as to what shadow would turn up next. And if it were not for this uncertainty, it wouldn't be a market. A predictable stock market is a contradiction in terms.

SUMMARY

This chapter is about second-order beliefs—beliefs about beliefs—such as are involved in the creation of institutions and thereby institutional facts ("i-facts"). It is put after the chapter on Hobbes because that philosopher's account of the creation of the State—the "Great LEVIATHAN"—is an early instance of attention to these issues.

J. L. Austin and John Searle are the chief discussants in the more recent literature. Natural facts are independent of beliefs; i-facts are not. Institutions, according to Searle, are created by *constitutive rules* expressing *collective intentions*. The formula is the speech-act "X counts as Y in context C," for example, this piece of wood counts as a knight in the game of chess.

Comparatively few important institutions, however, have been created by such explicit resolutions. They come about through tradition, legislation, claims of Rights, and the nature of the Economy. Their origins are easily forgotten, and indeed yet another Invisible Membrane hides their histories and makes them seem to be plain facts, perhaps of supernatural origin, like marriage. (It is hard to regard marriage as "a few words mumbled by a priest," though that is in fact what it is.)

All Rights are institutional. The concept of Rights, unknown to the ancients, seems to be of medieval origin, but political rather than theological, arising from the complexities of *homage* in the feudal system (or lack of system).

Theories of political obligation are either theological ("the powers that be are ordained of God") or Consent theories. The latter have always been based on a postulated condition ("state of nature") in which people who are all equal in the relevant respects (physical power, reasoning power) get together and agree on how they are to be governed. The Consent theory sketched in this chapter does not presuppose such unreal conditions.

The Python Effect, institutional collapse through loss of mutual confidence, poses a real, practical, and perhaps insoluble Problem of Induction: beliefs about beliefs cannot always be safely extrapolated.

CHAPTER 21

⌒⌣⌒

Miletus Preserved II: Spinoza

By decree of the angels and by the command of the holy men, we excommunicate, expel, curse and damn Baruch de Espinoza, with the consent of God, Blessed be He, and with the consent of the entire holy congregation.... Cursed be he by day and cursed be he by night; cursed be he when he lies down and cursed be he when he rises up. Cursed be he when he goes out and cursed be he when he comes in. The Lord will not spare him, but then the anger of the Lord and his jealousy shall smoke against that man, and all the curses that are written in this book shall lie upon him, and the Lord shall blot out his name from under heaven.... No one should communicate with him, neither in writing, nor accord him any favor, nor stay with him under the same roof, nor come within four cubits of his vicinity, nor shall he read any treatise composed or written by him.

{Nadler 1999, 120–21}

This proclamation was read out in the Synagogue of Amsterdam on July 27, 1656. Its object had offended the holy congregation by expressing opinions abhorrent to both Christians and Jews concerning immortality, the Holy Bible, the creation, democracy, and religious toleration. He was the target of an assassination attempt.

The principal work of this outcast, *Ethics Demonstrated in Geometrical Order,* is the greatest *single* work in philosophy. (I think this is a secure judgment. The only rival candidate would seem to be Plato's *Republic*. But the *Ethics* has the advantage—as I take it to be—that most of the statements in it are true.) It was published only after its author's death, and nearly not published at all, doing so being risky even in Holland, the freest nation in the world. Despite its author's profound learning in medieval Christian, Arabic, and Hebrew philosophy, it is at the modern end of the bridge we have been looking for that skips entirely the "medieval interlude" and preserves unmodified and clarified the three Milesian requirements for scientific Grand Theories.

This is evident even in its title and scheme of organization. Why *Ethics*—an odd and (to us) misleading designation for a Theory of Everything? Well, the ultimate aim of most, perhaps all, Greek philosophies was to describe and point the way to The Good Life—how to achieve genuine happiness insofar as it was in the individual's power. But a necessary condition of success in this endeavor (as in any) was knowing what you had to cope with—how things really are, what the true low beliefs about them are; hence the study of Physics, the science of Nature, as propaedeutic. For example, if you knew about atoms you would be relieved of superstitious fears. And you needed (what we now call) Psychology in order to know yourself and to deal more adequately with your fellows (even if they had excommunicated you). Furthermore, so Spinoza thought, since what you needed to know was also necessary in the sense of "what could not be otherwise," of which the truths of mathematics were the paradigm, the "geometrical order" was the most perspicacious way of expressing it.

Spinoza did not say within the body of the *Ethics* that the attainment of happiness was his supreme goal, but he did at the beginning of his earlier *Treatise on the Improvement of the Understanding*:

After experience had taught me that all the usual surroundings of social life are vain and futile; seeing that none of the objects of my fears contained in themselves anything either good or bad, except in so far as the mind is affected by them, I finally resolved to inquire whether there might be some real good having power to communicate itself, which would affect the mind singly, to the exclusion of all else: whether, in fact, there might be anything of which the discovery and attainment would enable me to enjoy continuous, supreme, and unending happiness. (Elwes translation.)

This good he found to be "knowledge of the union existing between the mind and the whole of nature"—an aspect of Unity—and in helping others to the same understanding.

Spinoza studied Descartes and wrote a book expounding his philosophy, with which, however, he profoundly disagreed. Descartes held that, strictly speaking, God was the only Substance, that is, Being capable of existence independent of everything else; but he did not speak strictly, allowing instead for the human being to consist of two substances, mental (thinking) and material (extended). How they could interact was then a problem. Spinoza held such interaction to be impossible.

What there is, and all that there is, according to Spinoza, is one Substance, its Attributes, and its Modes. Substance is also called God and Nature— God, because he (or it) is infinite, eternal, free, self-caused, and self-explanatory ("in itself and conceived through itself"). It is in no way a person (let alone three). Lacking nothing, it acts for no end (eliminating teleology), makes no choices; it is free because unconstrained, not because there is anything random in its action; on the contrary, everything that happens, happens as it does of necessity. Contingency names only human ignorance, nothing in God or Nature.

God, being infinite, has infinite Attributes ("what the intellect perceives as the essence of Substance"), of which, however, the human mind recognizes only two: Thought and Extension. Modes, some infinite, some finite, are the affections of Substance, those things that can only "be and be conceived through another." Each mode exists in every attribute: that is to say, it is both thinking and extended (etc.). So a given person—all of him or her—can be understood as a mind, and also—all of him—as a body. They are the same thing "expressed in different ways." Being identical, neither can act on the other. Events in the brain do not *cause* ideas (beliefs); they *are* ideas—quite unlike Descartes's "dumb pictures on a tablet," as Spinoza called them, they are active entities. In fact, they are beliefs; they carry affirmation with them. {See Matson 1994.}

God or Nature is the "immanent, not the transitive" cause of the modes. As we should say, God is energy (etymologically the word means "in-working"), not a Zeus-like pusher-puller. This is Spinoza's Naturalism or Immanentism, the most original and important of the Milesian insights. And all its workings are in strict accord with the rules that are not imposed upon but constitute the modes. It is the goal of natural science, both Physics and Psychology, to find out what these rules are; and, Spinoza assures us, the human mind is adequate to the task. This is Spinozistic Rationalism and Determinism. It implies, of course, that there are not and can be no miracles; expressing this was the feature of his philosophy most offensive to the popular religions.

Proposition 33 of *Ethics* Part I, On God, is, "Things could not have been produced by God in any other way or in any other order than is the case." Proposition 35 states that "Whatever we conceive to be within God's power necessarily exists." These emphatic denials of plural "possible worlds" and of so-called "logical possibility" express Spinoza's rigid Determinism and Rationalism.

It is easy to see from their integration into this philosophy how the three Milesian intuitions are not independent desiderata of a Grand Theory but intimately bound up with one another.

Now, how can understanding all this lead to the blessedness that Spinoza sought to attain and impart? Well, for one thing, since mind and body are identical, the body as such ought not to be neglected:

> It is the part of a wise man to refresh and invigorate himself with good food and drink, as also with perfumes, with the beauty of blossoming plants, with dress, music, sporting activities, theaters and the like, in which every man can indulge without harm to another. (*Ethics* 4 P45 Scholium)

But what a man—indeed every particular thing—essentially is, is the endeavor to continue in existence, to cope, as we have called it. And it is our ideas—our beliefs—that are central in the endeavor. These Spinoza divides into the adequate and inadequate. Adequacy is a matter of degree, the upper limit of which is the representation of its object and of its cause as it is in God—the clear and distinct low belief. Inadequate ideas are "fragmentary and confused." We cannot help having inadequate ideas, but the more we replace them with adequate ideas, the more our minds are active rather than battered about by the passive affects.

(Shirley in his translation renders—with apology—*affectus* as "emotion," but despite its unfamiliarity "affect" (White-Sterling) is better, as conveying the notion of outside influence. In Spinoza's day there was no English word "emotion," nor its equivalent in other European languages.) This is to say that Knowledge is Power, the power that really counts, to counter the passions, those confused ideas—beliefs—that keep us in bondage and make us miserable. Being parts of God or Nature, we cannot be free as God is, but recognizing the necessitation of everything and thereby coming to terms with it is the kind of freedom accessible to us. Spinoza even thought that since our minds are *composed* of ideas, to the extent that they are maximally adequate they share in God's eternity. {See Matson 1990.}

SUMMARY

Spinoza, more concerned than Hobbes with the ancient conception of the role of philosophy in delineating the Good Life, made Substance, God, and Nature into synonyms. God is eternal, free, and all-powerful, but in no way personal, operating for no end, but from the necessity of its nature. Nothing is contingent. This entity, of whose infinite Attributes we know two, Thought and Extension, and whose Modes are the particular things (including us) of our experience, is all there is. Mind and Body are "the same thing, expressed in two ways." A particular mind is composed of Ideas—beliefs, active entities, not the "dumb pictures on a tablet" of Descartes. Some ideas are adequate, others are inadequate, "confused and fragmentary." The more we replace our inadequate ideas by adequate ones, the closer we attain to blessedness, and indeed share in the eternity of God.

CHAPTER 22

✦

The Strange Case of David Hume

Beyond doubt David Hume is the eighteenth-century philosopher with the greatest continuing influence both in the Anglo-American sphere and, through his awakening of Immanuel Kant from "dogmatic slumbers," on the European continent as well. He it was who brought to attention as (allegedly) problematic Cause-Effect, Induction, Personal Identity, Knowledge of Other Minds, and the Existence of the External World; and while he did not originate the (alleged) problems about Free Will, the Existence of God, Material and Mental Substances, and Evil, he largely set the terms in which they continue to be wrangled. And he is the acknowledged patron of Skepticism and Positivism, the schools or attitudes of present-day philosophers who pride themselves on being progressive and hard-boiled; though some theologians, and philosophers sympathetic to them, are also grateful to him as one who (allegedly) showed that ultimately science rests on faith just as religion does, so the one is as reasonable as the other.

Hume did all of his important philosophy in his twenties, and (except for his notorious views on religion) published it in his three-volume *Treatise of Human Nature*, 1739–40, which "fell dead-born from the press, without so much as exciting a murmur among the zealots." When his zealot-baiting rewrite known as the *Enquiries* met a similar fate he gave up philosophy for history, becoming eventually the best-paid writer in Britain.

The views expressed in the *Treatise* do not add up to a Grand Theory, but they have been so influential on all subsequent such theories that we must discuss them in some detail. Furthermore, their primary aim closely overlaps that of the present work, being to give an account of why people have the basic beliefs that they do have.

Hume thought of himself as, or at any rate aspired to be, a scientist: hence, the subtitle of his *Treatise*, *"Being an Attempt to introduce the experimental Method of Reasoning into Moral Subjects."* But it is not easy to find any experimental reasoning in this work, especially in Book I: Of the Understanding, where the fundamental philosophy lies. Actually his method was the same as that of Hobbes: introspection plus observation of human behavior. Unlike Hobbes, however, Hume lays down on his first page a succinct but comprehensive theory of what there is to investigate in "Moral Subjects," that is, what we nowadays call psychology and the social sciences. The human mind consists entirely of "perceptions," which

> resolve themselves into two distinct kinds, which I shall call IMPRESSIONS and IDEAS. The difference betwixt these consists in the degrees of force and liveliness with which they strike upon the mind, and make their way into our thought or consciousness. Those perceptions, which enter with most force and violence, we may name *impressions*; and under this name I comprehend all our sensations, passions and emotions, as they make their first appearance in the soul. By *ideas* I mean the faint images of these in thinking and reasoning; such as, for instance, are all the perceptions excited by the present discourse, excepting only, those which arise from the sight and touch, and excepting the immediate pleasure or uneasiness it may occasion.

Hume thus takes unargued as his starting-point the "new way of ideas" that John Locke had derived from Descartes. Hume's "perception" is the same as Locke's "idea," which is "whatever it is which the mind can be employed about in thinking." That is to say, what we have "before the mind," on which they "operate," are not shoes and ships and sealing wax but in-between-things, whose relations of resemblance and causation to the actual shoes etc. (if there are any) are problematical.

Both Locke and Hume were far from being avowed Cartesians. Locke indeed began his enormously influential *Essay Concerning Human Understanding*—the acknowledged foundation piece of The Enlightenment—with a tedious polemic against Descartes's theory of "innate ideas"; but he did not question the notion of the idea as such. Likewise, Hume dismissed Cartesian doubt with the remark that "were it ever possible to be attained by any human creature (as it plainly is not) [it] would be entirely incurable" {*Enquiry* 116}, without, however, noting the dependency of the "idea" on Descartes's toying with global doubt, superficial though it was. Probably he was thinking of Descartes's forgetting to dare the Evil Demon to deceive him about logical and causal principles. He moreover takes for granted that serious philosophy must be inside-out, that is, must begin with what is (allegedly) known with certainty, to wit, the contents of the philosopher's mind, and build on those "foundations" his theory of what the world out there (if any) is like. (Among the ancients only Plato's philosophy, that of the Cyrenaics, and Skepticism had been inside-out.)

Hume adopted the "perceptions" theory because he believed that it was true. In any case, by the 1730s it was the only philosophical game in town. No one except "the vulgar"—people ignorant of the most basic and obvious philosophy—any longer disputed that what we *directly* perceive are not (physical) lawns and lemons but (mental) green patches and sour tastes. {On this tricky word *directly*, a favorite of Bishop Berkeley's, see Stroll 2009, chapter 2.}

Hume next sides with Locke against "innate" ideas, laying it down as a first principle that the mind cannot create a simple idea. Every idea, or at any rate its components, must be copied from an impression. This is Empiricism. Validating an alleged concept as meaningful then consists in producing the impression(s) that gave rise to it.

It is hardly surprising that with this limited apparatus Hume has difficulty in explicating many basic notions having to do with human knowledge, for example, belief, which he defines as "A LIVELY IDEA RELATED TO OR ASSOCIATED WITH A PRESENT IMPRESSION," defending this conception with the dubious claim that we get more excited when we read a story as a true history than when we regard it as mere fiction. However, we need concern ourselves only with what he has to say about the processes whereby we acquire knowledge.

The truths at which we aim are of only two kinds: relations of ideas, and matters of fact. The former are virtually restricted to mathematics, the science of quantity or number. Here we can reason a priori and with certainty (at least in arithmetic and algebra. Hume had some qualms about geometry, the science of space). Matters of fact and existence must be dealt with a posteriori, by experimental reasoning which can yield only probable conclusions. The central relation in such investigations is that of cause and effect, for the scrutiny of which Hume was best known.

It was about time. Descartes, as we have noted, without discussion exempted cause-effect from the domain of possible evil demon deception and relied on it entirely and uncritically to get himself out of the solipsistic slough in which he found himself at the end of the second *Meditation*. His skyhook, the proof of the existence of God, is that he finds in himself the idea of God, an infinite and perfect being; this idea must have a cause; and "it is manifest by the natural light that there must at least be as much reality in the efficient and total cause as in its effect....And this is not only evidently true of those effects which possess actual or formal reality, but also of the ideas in which we consider merely what is termed objective reality." So, since Descartes himself doesn't fill the bill, the cause of this idea of God can be none other than God Himself. (One may wonder how a man who could doubt that $2 + 3 = 5$ should have declared himself unable to doubt this whopper. Perhaps he conceived of cause/effect as being a purely logical relation.)

Descartes bequeathed to his successors another big cause/effect worry, that of how thinking unextended substance could interact with unthinking extended substance. His tentative solution, that the soul rerouted the (material) animal spirits from afferent to efferent nerves in the "pineal gland" in the midst of the brain appeared unsatisfactory to the Princess Elizabeth of Bohemia and ultimately to Descartes himself. It is hard not to think of other attempts to solve this puzzle as simply crazy. The Occasionalists held that on the "occasion" that a baseball bat hits one's head, God (not the bat) causes the soul to feel pain. Malebranche extended this view to all alleged causation: God is the only cause there is. And Bishop Berkeley took the ultimate step: matter, not causing anything, doesn't even exist.

Hume prudently avoided involvement in these controversies by saying that perceptions arise in the soul "from unknown causes." He asked not what causes what, but what is *meant* by statements of the form "A causes B," and why people believe them.

Hume defines "cause" in two ways. Both begin with the specification that "A CAUSE is an object precedent and contiguous to another." The first definition adds to this "and where all the objects resembling the former are plac'd in like relations of precedency and contiguity to those objects, that resemble the latter." The second adds instead "and so united with it, that the idea of the one determines the mind to form the idea of the other, and the impression of the one to form a more lively idea of the other." These are not equivalent: the first describes what it is for object A to be the cause of object B (whatever "objects" are)—namely, that A be before and close to B, and that all objects "resembling" A are before and close to objects after and close to objects resembling B; whereas the second is in terms of the (causal!) effects of these juxtapositions on the mind of the spectator. But passing over this difficulty, let us note that Hume's description of the perception of the causal relation is essentially identical to our account of how a low belief—an expectation—is formed. However, we are disappointed that the author, instead of proceeding from this beginning to show how beliefs about causal relations can be combined, as in Miletus, to advance to the level of science—principally through the relation of "tethering," which has no equivalent in the Humean philosophy—he instead devotes his attention to debunking the contention that there is anything more (or at any rate, anything more that can be known) about the causal relation than constant conjunction. Occasionally he refers to "secret springs" which, for instance, make bread fit nourishment for the human body, but always he pessimistically declares them not just unknown but unknowable. (It is odd to think of nutritional scientists as refuting Hume, but they do.)

Hume recognizes that people think there is more to cause and effect than constant conjunction: there is an idea of "necessary connexion." A not only *is* but *must be* followed by B. This idea, like all ideas, must be the copy of an impression; but what is that impression? From reason, one might think. But reason, for Hume, is limited in its efficacy to "comparison of ideas," as in—and only in—mathematics. At length he concludes that the impression in question can only be the *feeling*, generated by many repetitions, that when an A occurs, a B is in the offing. So necessity is not "out there" but "in here," though it is human nature to project it on to the "objects." It is "custom operating upon the imagination."

An essential premise of Hume's argument for this conclusion is that if reason could establish the *must* relation, it would have to show that every alternative was *absolutely impossible.* But this it can never do, for we can always *imagine* an alternative outcome to A's usual attendant—we can picture the second billiard ball when hit by the first as remaining motionless, or exploding, or whatever—and nothing that we can form a clear idea of can be absolutely (i.e., logically) impossible. At this point the sympathetic reader of this book (if there is one) should have the "Aha!" experience. Here again is the Cartesian hangover from theology, the contingency of the world.

This is not the last place where the Evil Demon peeks out at us from the Humean philosophy. His most notable appearance is in connection with Hume's celebrated Problem of Induction. (Hume didn't call it that, or give it any name.) What warrants inferences from perceived objects to unperceived? The argument "All perceived As have been conjoined with Bs, therefore a B will come after the next A" is invalid as it stands. If Reason is to establish its conclusion, another premise must be added, to the effect that "the future will resemble the past in the relevant respect": a Principle of Induction. But, Hume avers, this principle is neither self-evident nor provable by Reason, because (of course!) "We can at least conceive a change in the course of nature; which sufficiently proves, that such a change is not absolutely impossible." And to try to establish it inductively would be obviously going in a circle. Nor would adding "probably" help. So there is and can be *no reason* for our belief in the validity of induction.

'Tis not solely in poetry and music, we must follow our taste and sentiment, but likewise in philosophy. When I am convinc'd of any principle, 'tis only an idea, which strikes more strongly upon me. When I give the preference to one set of arguments above another, I do nothing but decide from my feeling concerning the superiority of their influence. Objects have no discoverable connexion together; nor is it from any other principle but custom operating upon the imagination, that we can draw any inference from the appearance of one to the existence of another.

Tell that to the sublime Mr. Newton sitting under the apple tree and getting conked on the head, one is tempted to suggest. But the curious thing is that Hume has something like the same opinion of his own official philosophy. Again and again skeptical pronouncements are interspersed with remarks such as "'tis in vain to ask, *Whether there be body or not?*" for "Nature has not left [belief in continuously existing bodies] to ... choice, and has doubtless esteem'd it an affair of too great importance to be trusted to our uncertain reasonings and speculations." And he even recommends "carelessness and inattention" as antidote to his own philosophy!

It is hardly necessary to observe that the Problem of Induction vanishes once the notion of "logical possibility" as a kind of real possibility is given up; and that the Problems of Personal Identity, Mind/Body, Other Minds, and Knowledge of the External World likewise fade out (or can be consigned to sciences such as brain physiology) once the notion is abandoned that the Foundations of Knowledge must lie in solipsistic subjectivity. Yet these puerile topics persist in colleges as staples of Philosophy 101, the lecturer having to pretend that there is something almost ineffably profound about them. This is a disgrace to philosophy, which, whatever else it may be, should deal with first and last things.

It does not matter when a clown like Bishop Berkeley, bemused by Cartesianism, fills his pages with absurdities. But Hume! the most intelligent man in Great Britain, entirely free of superstitions and bravely exposing them to the world when it was still personally dangerous to do so, and (as we shall see) so right about "Morals," expounding "principles" which if taken seriously would have strangled science! One cannot read his writings on Understanding without a profound melancholy at so great a figure so gone astray. Let us read again the famous concluding paragraph of his *Enquiry Concerning Human Understanding*:

> When we run over libraries, persuaded of these principles, what havoc must we make? If we take in our hand any volume; of divinity or school metaphysics, for instance; let us ask, *Does it contain any abstract reasoning concerning quantity or number?* No. *Does it contain any sexperimental reasoning concerning matter of fact and existence?* No. Commit it then to the flames; for it can contain nothing but sophistry and illusion.

Yes, Hume's principles do justify that modest condemnation. But what an irony, that those principles themselves are not self-evident or even true but derive crucially from the "divinity and school metaphysics" of the Omnipotent Creator-Legislator, the virus that Descartes left to infect the human computer.

Norman Kemp Smith's widely accepted conjecture that Hume wrote Book III: Of Morals *before* Book I of his *Treatise* conveniently explains why

Hume's moral theory is so much more satisfactory than his theory of knowledge; for in writing of why we make the moral judgments we do, Hume is not constrained by the Cartesian rigmarole of Ideas but can base everything on his native wit. His well-argued thesis, that moral judgments are based not on Reason but on feelings—"passions," emotions—instinctive to social animals, is in keeping with what we might infer from the theory of high and low beliefs, which furthermore can explain why the "monkish virtues" based on religion are in reality vices, as Hume maintained. It seems appropriate then to devote the next chapter of this book to that theory.

SUMMARY

David Hume had the most lasting influence of any eighteenth-century philosopher. Almost all of the standard "Problems of Philosophy" owe their origin or at least the terms of their discussion to him. So we need to consider him carefully even though he had no Grand Theory.

His account of the human understanding begins with blanket and unargued acceptance of the Cartesian inside-out approach: we must start with our "perceptions" and build knowledge on their "foundation." Perceptions are either impressions—sensations and feelings—or ideas, which are their faint copies. The mind cannot create any simple idea; so the validity of any idea can be established only by tracing it to the impression from which it originated. Applying this principle to our idea of cause and effect, we find that it is derived from experiences of objects contiguous in space, precedent one to the other in time, and constantly conjoined. But we have besides the idea that cause and effect must be "necessarily connected." This idea can only arise from the feeling that we get, after experiencing many instances of the sequence, of expectation that if an A appears then a B will follow. The feeling is in us, but we project it on to the world.

Can reason validate inferences from the observed to what is not observed? No. The argument "All observed As have been followed by Bs, therefore the next A will be followed by another B" is logically invalid; it needs to be supplemented by a "principle of induction," that the course of nature will continue with regularity. But Reason cannot prove this, for a change in the course of nature is conceivable, and nothing conceivable can be "absolutely impossible." So there is *no reason* to believe that the future will resemble the past, Hume feels forced to declare. But once this move is recognized as just another application of the medieval Contingency of the World principle again, there is no more problem.

In his consideration of Morals, however, where his thought is not in thrall to the Evil Demon, Hume reasons justly to a theory of judgments rightly based on feelings.

CHAPTER 23

✧

Ethics Without Edification

One would like to be able to say that mankind have at last "grown up," that we have no further need of social high beliefs, which in any case cannot be sustained as beliefs when they are recognized as high. It seems to be possible to attain, or at least to approximate, the condition of having eliminated high beliefs from one's view of how things are. That is another way of expressing the Lockean ideal of proportioning the degree of one's assent to the weight of evidence.

Low beliefs, however, are seldom edifying, in the sense we have given that word, of promoting the viability of the social group. If high beliefs are expunged, what happens, then, to morality, conceived as rules whose general observance is a necessary condition for keeping a society from disintegrating?

The question concerns behavior. There is little difficulty in showing that certain kinds of actions benefit, and others harm, the community of which the agent is a member; and that the integrity of the community is (usually) in the interest of all its members including the agent. What, specifically, these actions are, is a matter for empirical determination. If, then, morality is simply the general word for acting in the communal interest, it can be shown—from low beliefs—that moral behavior is in everyone's interest. As from the standpoint of action the rational person is usually defined as one who acts on the best evidence so as to maximize his or her expectation of benefit, morality is rational, and thus there is no problem, in theory at any rate.

There are, alas, some age-old catches. For one, if the public interest coincides with the private, it does so in the long run, but we have to live in the short run. And the public interest is only one interest, which may be and often is overridden. The rational legislator knows that taking a bribe weakens the community, but on the other hand it may be her only way out of a ruinous financial situation. Then there is the "prisoner's dilemma": action in the public interest is often effective only if the agent can expect everybody else to behave the same way. If we all install expensive smog reducers, we will all benefit; but if some do and some don't, those who do lose.

These are some of the problems to which high beliefs provide a solution (of sorts) by assuring us that we do not live only in the short run, and that we cannot hope to escape the surveillance of supernatural powers who punish those who put their own interest ahead of the community's. However, the effectiveness of high beliefs in this regard is limited, and no society ever relied on them exclusively. The sanction of public opinion, in the hunter-gatherer band, together with formal laws and enforcement procedures in larger communities, has always done most of the work. High beliefs do not have so much effect on behavior by directly enjoining or forbidding certain actions, as they do in promoting a sense of community, from which communally beneficial action flows, as it were, automatically. It is the weakening and even disappearance of this spirit that is the most deleterious effect of the decline of high belief.

Let us not forget, however, that there is a morality built into any gregarious animal by evolution and which therefore antedates language and high beliefs. The content of this "natural law" is the set of behavioral biases required to keep any herd, pride, gaggle, or band from splitting up because its members think they could get along better on their own. Mainly it is concerned with internal peace (or at least limits on aggression), cooperation in warding off predators, getting food and sharing it (to some extent) when it is got, protecting the young and the helpless, and regulating sexual relations. These add up to "community spirit" in the sense that the animals concerned behave (for the most part) in accordance with these norms within their group but differently when interacting with outsiders.

This is enough to show that there is no *conceptual* dependence of morality on high beliefs. Our question, however, is the *factual* one, whether morality as it exists in our time is likely to be able to survive the extinction of the high beliefs that now and in the recent past have given it support.

Let us consider the question using the Lockean "historical plain method" that we have hitherto employed with regard to beliefs in general.

To reiterate some basic points: In animals of much complexity, not all actions will be reflexive or instinctive. More than one possible action may suggest itself as appropriate, and the animal must *choose*. Perception results in *belief* on the part of the animal as to how things *are*. Subsequent action expresses the animal's *choice*: in light of the belief, this is what the animal *ought* to do. "Ought" is used here in the broadest sense: all things considered, this action will best promote coping. Perceptually induced beliefs are *true* when the lay of things is as perceived, that is to say, when the animal's expectations are (or would be) borne out in the event, so that its effort to cope is not frustrated by misapprehension of the facts of the situation; otherwise such beliefs are *false*. Choices are *right* when the animal's action is more apt for coping than the alternatives; otherwise they are *wrong*.

The true/false distinction is much sharper than the right/wrong. Since beliefs are less detailed than the states of affairs they purport to represent, and elements of approximation always enter, the relation of a true belief to its object is not one of strict correspondence. Nevertheless, a somewhat fuzzy "law of excluded middle" holds for perceptually induced beliefs; that is, it is possible to tell (at least after the fact) whether the belief was true or false. No such clarity reigns in the realm of right and wrong. Even if the action was successful, some other might have been better; and even if its result was calamitous, it still might have been the best that could be done. This means that in making our choices of the right thing to do we must rely on statistics, as it were; we judge by whether the proposed action conforms to a *rule* the observance of which has been shown by experience to lead, on the whole, to better results than contrary behavior probably would.

Reflexes and instincts are wired-in rules of this sort. Evolution does the wiring. Thus, reflexive and instinctive behavior is right far more often than it is wrong—practically always, if the circumstances are similar enough to those in which the instinct evolved.

Reflexes and instincts by themselves seem to be enough to keep ants and spiders going; but for coping with novel situations, an ability to *learn how* is essential. Learning-how consists in progressive modification of the perception-to-will nexus in the light of *memories* of (i.e., stored beliefs about) previous similar situations. The outcome of successful learning is more efficient coping, acquisition of facility in making right choices in a broad variety of situation: new *precepts* of action.

Evolution has honed instincts and perceptual and learning abilities to the point where the animal can be counted on to do its best in most situations that it is likely to encounter; that is to say, it is "adjusted." Collisions with asteroids, and incursions of human beings with panoplies of guns, traps, and pesticides are events that it cannot cope with, but that is not the beast's fault, nor evolution's either.

GREGARIOUS ANIMALS

For gregariousness to be an advantage, behavior within the community must be such that the probability is fairly high of an individual's being better off, that is, coping better, in it than out of it. This means, first, that security must be enhanced by membership: the group must be more effective at warding off predators than the individuals separately would be; moreover, intra-communal aggression must be, if not eliminated, at any rate less baneful on the whole than what the members would experience on their own. Peace must be kept. From this requirement a number of fairly definite characteristics of gregarious behavior follow:

1. There will be a hierarchy (pecking order), for if there is not, if individuals do not know their places vis-à-vis the others, there will be constant bickering or worse. To be sure, the hierarchical order often gets established by violent competition; but that process must be kept within bounds (recognized "I give up" signals, etc.).
2. The young must be nourished and protected.
3. The members of the group must aid each other, offensively and defensively, to the extent feasible by animals of their kind.
4. Surplus food must be shared; at least there must be some rule of precedence in feeding, and no individual, however high in the hierarchy, can monopolize the supply.
5. Communal outrage must be directed at members of the group believed to have transgressed these precepts. It may manifest itself in various hostile acts. This is the germ of justice and punishment.

These are some elements of *morality*, that is, the structure of biases on the behavioral choices that individual members of a gregarious animal species make in their dealings with fellow members of their community. Evolutionary processes will result in building into the genes tendencies to behave in the right way—"right" meaning "promoting survival."

These rules apply within the community, but not necessarily between communities. However, inasmuch as biological fitness requires exogamy, there will be a sort of "Geneva convention" between different communities of the same species, which will not, in general, indulge in no-holds-barred aggression toward one another as they do with prey species. (Unfortunately, *Homo sapiens* sometimes tends in the direction of being an exception to this rule.)

The individual animal, hungry but refraining from eating its den mate, may not experience its forbearance as a restraint, or, indeed, have any conception of itself as an individual, or any other conceptions for that matter. It is nevertheless the case that the animal's behavior, in the social context, is biased in the direction of furthering the group interest (as defined by survival). That is enough to constitute the behavior as proto-morality. It is of course in no way conventional.

LANGUAGE

No doubt the proto-moral rules are more numerous and more complex for chimpanzee society than for (say) zebras. But the enormous divide between one (viz., our) species of gregarious animal, and all the others, is occasioned by the acquisition of language.

All social animals can communicate with each other to some extent; but only ability to manipulate symbols makes possible the notion of the future and the conception of present action as having a bearing on what is to come. An offshoot of that notion is the *promise* and associated concepts such as trust and duty. A betrayed trust, a shirked duty being socially disruptive, further moral rules come into being. These new rules can be and are explicitly taught: children can be told what is expected of them rather than having to find out by trial and error.

Without language it is difficult or impossible for one animal to get another to do some specific thing. With language, *orders* (and requests) become possible. This gives a new dimension to social activity, which now becomes *ordered*—on the basis, to be sure, of an already existing hierarchy. The virtues of obedience and loyalty, and the vices of insubordination and treason, come into being.

LOW MORALITY

The root principle of animal association is that internal peace must be established and kept: aggression between members of the group must not be allowed to proceed to annihilation of the losing party. Earlier we listed minimal conditions of this sort which must obtain in any successful group of gregarious animals: we called this the herd's morality. To the list we may add, for human beings,

6. Truth-telling.
7. Promise-keeping.

Let us call this set of internal constraints on the interpersonal behavior of human beings *low morality*. It is continuous with the morality of subhuman gregarious animals and is basically the same for all human groups, being a specification of how "normal" people must habitually behave if a community is to be viable—though precisely how the requirements are to be realized will vary with the size, complexity, and capabilities of the particular group. For human self-sufficient communities at the current level of technology, experience seems to show that unless the following conditions are substantially met, intra-group dissatisfaction may rise to a dangerous level, that is, there will be a widely shared perception that one would be better off living entirely on one's own—"in the state of nature," they would have said in the seventeenth century—than in the community.

1. Who has control of whom and what—that is, power and property—must be known attributes of persons, and alterations must proceed in regular ways, recognized as in accord with perceptions of entitlement, including merit, not haphazardly or by sheer force. This is the chief element in the rule of law: people must be able to count on obtaining and retaining the status to which they are entitled.

2. Although primary responsibility for care of the young will in general rest with their parents, the community will take over if necessary. Care will include not only nourishment and protection from attacks and other hazards, but education to a level appropriate to the child's status and potential.

3. Members of the community must be liable, according to their abilities, to assist in communal defense, and to aid each other in calamities.

4. Those unable to provide necessities for themselves must be cared for communally (a social safety net). Communal costs must be shared; those more able to pay must pay more—not, however, to the point of confiscation, which would be incompatible with requirement (1).

5. Wrongdoers—perpetrators of *mala in se* (acts wrong in themselves), and of *mala prohibita* (acts which are made offenses by positive law) insofar as the prohibitions are socially necessary—must not be allowed to profit from their offenses, but must suffer socially imposed penalties expressing the community's disapproval and serving as deterrents.

6. The things that people say must be correlated with what they believe. Otherwise, communication—a social necessity—becomes impossible.

(Of course, much fibbing is tolerated and even required in every community. I have in mind only Donald Davidson's—and Kant's—point that if lying becomes the rule rather than the exception, communication breaks down. Religious and political discourses present special problems in this regard.)

7. One must be held morally delinquent who does not, in general, keep his or her promises. Human society is largely structured on the basis of trust.

The low morality, it should be noted, is concerned with the *right* (thing to do) and not with The Good, some shining ideal condition to the realization of which individual or social effort must be directed. The Good is essentially a religious or political concept, not a moral one. Low morality is not concerned with ultimate aims, as long as they are not incompatible with doing the right thing; and this leaves much leeway. As far as low morality goes, it is all right to pursue wealth, or learning, or political power, or pleasure, or stamp collecting as long as the means employed are aboveboard. The "moral ought" may be characterized as what is the right thing for an individual to do, as a member of a community, with respect to actions affecting the interests of other members of his or her community. These types of actions (and abstentions) may be spoken of as *duties.*

Children in a band, the hunter-gatherer group of forty adults, more or less, are taught the right things to do with respect to building shelters, making and shooting arrows, hunting—and dealing with their fellows. The social skills are skills like any other: they are learned by precept and example; some are better at them than others, but nearly all are pretty good; excellence is praised and delinquent performance criticized. A reflective, even philosophical, tribesman would not be likely to see in them any mysterious quality setting them apart from other desirable accomplishments. Social obligations are not experienced as burdens imposed from outside, or as in need of justification. The question "Why be moral?"—if *per impossibile* it were raised—would seem as odd as "Why shoot straight?"

Nevertheless, that question would have an answer, though the tribesman would not know it. It would be, This kind of behavior is in fact a necessary condition of a viable community; consequently, will and aptitude to it have been built into your genes in the evolutionary process. It is part of your nature; you would, literally, not be a human being if you lacked it. This is substantially the answer of Protagoras in Plato's dialogue of that name.

Low morality persisted through the invention of agriculture, augmented as, in consequence of that invention, social relations became more complex. Nevertheless, it remained universal, and does so to this day, for the necessary conditions for a viable society are everywhere the same. *There is no alternative to low morality.*

IMAGINATION

Although *low* morality is constant and transcends tribal boundaries, it is only the common core of actual comprehensive moralities, which of course exhibit much variety. The variable element I shall call *high* morality. To explain it, let us return to consideration of the anthropoid group that has just acquired language.

Language creates *imagination,* first in the basic and literal sense of ability to summon up images "in the mind's eye." With words, reference can be made to things not present where or when the utterance is made. When the narrator relates what has happened on the other side of the mountain, the auditor forms a "picture" of what it must have been like, without having actually been there. This is something that an animal without language cannot do, at least not on cue. A faculty of creating images of what is or was, immediately leads to imagination in the more usual sense, on "envisioning" what is not and was not, and perhaps never will be—that is to say, the contents of stories and lies.

Belief is liberated from its former necessary connection to perception: one can believe what one is told, and if the belief is not of such a kind as to be shown up as false by immediate confrontation with experience, it may persist indefinitely, for example, belief that some remnant of the individual consciousness migrates to another world after bodily death. We have called such beliefs, that have not *in fact* been subjected to the risk of contradiction by experience, *high* beliefs.

Inasmuch as high beliefs originate in stories, and all stories are about people, their content will be built on analogies to human beings and their ordinary experiences. It is characteristic of us that when things go wrong, we impute *guilt* to persons whose actions are supposed to have brought about the mishaps. Imagination makes possible the envisaging of superhuman yet personal powers to which guilt can be ascribed for whatever is extraordinary, mysterious, and important for well-being—weather, famine, earthquakes, tempests, fertility, and all the host of things and events on which human existence is dependent but which are beyond the primitive human being's control. This assignment of guilt is the simplest kind of causal explanation.

RELIGION

The idea is not long in coming that these powerful persons may be amenable to some of the influences that serve to modify behavior of one's fellows: threats, entreaties, flattery, promises of reward. Such stratagems will be tried, and sometimes they will "work". *Religion* gets established. Religion is a fabric of high beliefs, and of actions predicated on those beliefs.

If this account is correct in outline, it is already clear that morality as such cannot be dependent on religion. Morality exists wherever there is a herd; but religion depends on imagination, which in turn depends on language, which itself presupposes society; so the institution of religion in even the most rudimentary form must postdate that of morality. Nor is it the case that only religion can provide motives for moral behavior, since rules of morality are (sometimes) followed by beings without language and therefore without religion.

This is not to say, however, that religion has nothing to do with morality. Religious beliefs and other high beliefs have various functions, one of which is to make sense of the world. If the important uncontrollable things are thought of as manifesting the powers of supernatural hominids, that means they come within the ambit of the categories of human thought and desire, such as contrivance, love, fear, will to power; and to that extent their strangeness is mitigated and hope achieves a foothold. Earthquake and drought cannot be coped with, but perhaps Poseidon and Zeus can.

This imaginative view of things personalizes the world. It is misleading, however, to express this fact by saying that to the primitive religious mind the world is a "Thou," or even a multitude of thous. True, the gods are objects of possible communication; but they are seldom thought of as members of the community. Typically, the things thought to need explaining are calamities; the forces controlling them, therefore, are modeled on "Them": they tend to be supernatural counterparts of the menacing hostile tribesmen lurking in the interior of the forest. These powers are often thought of as engaging in practices that would be grossly immoral if done by people, for example, incest. This is just what should be expected, for religion is *essentially* concerned with the uncontrollable and antisocial forces that run things.

It becomes a high-priority community project, then, to placate these forces. The usual means for doing so is sacrifice—giving (or at least promising) precious things to the gods, as one might buy off hostile tribesmen. Now, in such transactions morality is not involved in any positive way, any more than it is in relations with alien tribes. Nor are the gods conceived as moral agents; to the contrary, the reason it is so hard to deal with them is their perceived malice. As part of the process of coping with them, the strategy may be adopted of making them "honorary tribesmen," as it were— but if this is done, the aim is to put them in a position of sharing in the tribe's moral concerns. No one dreams of them as constituting the *sources* of morality.

Indeed, the dangerousness of the supernatural is likely to have a negative effect on morals, in that the need to appease the gods may *trump* ordinary moral values. Deliberate killing of a blameless fellow is always forbidden by the low morality; nevertheless, it may be deemed necessary to kill someone—even someone important, as the Chief, or beloved, as the Chief's daughter—in order to placate the powers. "Teleological suspension of the ethical," to use Kierkegaard's euphemism {see Kierkegaard 1941, page 64} is the rule rather than the exception in primitive religion. Practices intended to appease supernatural forces will be ritualized, that is, will become standard and stereotyped; and ritual will take precedence over low moral duty.

HIGH MORALITY

Further, since the modi operandi of the gods, as well as their names, natures, sexes, and other attributes, are all products of imagination, the beliefs and rituals about them will vary more or less at random from tribe to tribe. That is to say, high beliefs and the practices based on them, unlike low beliefs and practices, will not be uniform; what is believed and done on this side of the mountain and on the other side will be "diverse." *We* want to say that this shows the rituals to be conventional. To the people in the actual situation, however—especially when, before the invention of writing, a practice can attain the status of having been what was done "from time immemorial" in a few generations—the rituals will be seen and felt as "the right thing to do" in just the same way as the right practices enjoined by the low morality. They may be distinguished as "piety," but the important thing is that one who does not conform to them is regarded as "bad," deficient in the qualities necessary for social living.

These new duties and obligations constitute the *high morality*. The actual morality at a given time and place, then, will be a compound of high and low. Hence, despite the common core of low morality shared with all the moralities of all the other tribes, a particular moral code considered as a whole will be perceived as quite different from that practiced elsewhere, and in consequence the natural view that strangers—"Them"—are "immoral" will be reinforced. For actions (e.g., burning babies to death) that in themselves would be considered wrong according to the low morality may, if believed to be required by the gods, trump that low morality and thereby take their place within the high—and therefore in the total—morality.

The upshot is that the way is paved for confrontations like the famous sociological experiment performed at the court of Persia:

Darius [the Great King of Persia] called together the Greeks who were around and asked them what price they would demand for eating the bodies of their dead fathers. They said they would not do that for any amount. Then he summoned the Indians called Callatians, who eat their parents, and asked them—in the presence of the Greeks who knew through interpreters what was being said—how much they would want for burning their fathers' remains. Loudly shouting, they bade him not to talk about such things. {Herodotus III 38}

Observations of this sort in turn lead to the belief that morals are "relative." In a way they are; but only in a way. What is really plural is religion (and the *high* morality derived from it), not the rules as to how it is necessary to behave if a society—any society—is to be viable (*low* morality).

COMMANDMENTS

It would be hard to estimate the amount of harm done by the assumption that moral (and legal) prescriptions have the character of commands.

—*Robert Fogelin*

How gods are related to the peoples who acknowledge them is a theme with many variations. They may, as we have pointed out, be conceived as demanding or hostile, to be sated or appeased at enormous expense (Baal, Huiztilopochtli, Homeric Apollo), or at any intermediate position in a more than one-dimensional continuum. Seldom, however, are they conceived as actual leaders of the people, or as members of the community within which the low morality is operative.

With one big exception: among the Jews. Yahweh came to be thought of as the generalissimo of the Hebrew people, personally smiting their enemies. As leader, he demanded obedience and punished insubordination. The novelty of this notion—that a god might tell you what to do, and it would be sinful (immoral) not to do it—has not been sufficiently remarked. It is virtually absent from Greek religion. The gods gave advice, and it might be folly to ignore it (only "might be"—on numerous occasions the gods engaged in deceit), but it was not sin or wickedness to do so.

In the Hebrew literature, on the other hand, the first sin was disobedience—the flouting, moreover, of a quite arbitrary prohibition. There is nothing in low morality that makes it immoral to eat fruit of any kind. In the book of *Exodus* the doctrine is developed in such a way that both high and low morality are construed as divine commandments. This is extraordinary. It is inconceivable that even the benighted Egyptians were unaware of the necessity, in any viable society, of discouraging murder, theft, adultery, perjury, envy, and neglect of aged parents. But these rules, part of the universal low morality, are made by Moses into the Second Table of Yahweh's commands, while the principal Jewish rituals and taboos, their high morality, are codified in the First.

This episode marks also the birth of the very first moral *theory*. As in the television game show *Jeopardy!* the answer comes before the question. The answer is, Yahweh has commanded His chosen people to behave morally, and will punish disobedience. The question is, Why be moral?

Pre-*Exodus* morality stood on its own feet, the way rules for adding numbers or firing pots do: this is simply the *right way*. However often and flagrantly the moral precepts might have been flouted, they were nevertheless recognized as carrying their own obligation. You shouldn't go around murdering and lying because murder and mendacity are *wrong*. You will, further, suffer unpleasant consequences if caught murdering and lying; but that is not what makes murder and perjury wrong; they simply are. Or perhaps the Egyptian mother would rhetorically ask her errant son, "What if everybody did that?"

Moses changed all that. You were to abstain from murder and perjury from "fear of the Lord," fear that he would smite you, or even the whole people, if he were disobeyed. Thus, disobedience, for the Hebrews, became the only vice (sin), and avoidance of divine punishment the only moral motive. Nietzsche was partly right: the Jews did bring about a fundamental revolution in morality; but it was not the one he described in *Beyond Good and Evil* and *The Genealogy of Morals*. Rather, it was this assimilation of morals—*all* morals—to commandments.

Of course, Moses did not do this in one fell swoop. Recalling our distinction between high and low moralities, it is evident that any high-moral precept is in effect a divine commandment, either directly and explicitly, or a priestly strategy for securing divine favor, which amounts to the same thing. In other words, the divine-command theory *is* true of what we have called "high morality." Moses's innovation was in conceiving the low morality (the Second Table) as having the same kind of status; and this would have been very natural for a tribe that already untypically held its sole god to be both commander and father of the people.

LAW, MORALITY, COMMAND

But while low morality is a system of constraints on behavior that can be expressed (to some extent) in precepts and rules, it is not and cannot be a set of positive laws or commands.

Laws are not commands; however, we need not go into their differences here. For both laws and commands presuppose a conception foreign to low morality: that of an external authority capable of imposing a sanction for noncompliance. Commands (from individuals) and laws (from legislatures) are speech acts that are infelicitous unless the utterer is a person (natural or corporate) invested with competent authority to initiate punishment for disobedience. Further, commander and subordinate, law-giver and subject, are distinct persons.

None of these features holds in the relation between morality and the moral person, unless metaphorically. The requirement of low morality is not experienced as imposed from without, much less as expressing the will of an external authority. It is more analogous to aesthetic judgment. Guilt or remorse does not feel like the consequence of having disobeyed an authority or of having failed to carry out a command; it is, rather, self-loathing for not living up to the model of humanity that one has put before oneself. This at any rate is how Aristotle and Spinoza looked at the matter, though not how Saint Paul and Saint Augustine did.

But not only does the concept of command involve these three notions—authority, penalty, and externality—that are foreign to low morality; it is also the case that command lacks the basic moral character of *normativity.* Anything that it is possible to do can be commanded, and its being good, bad, indifferent, virtuous, or wicked has no effect on its status as command. To equate moral rules to commands is therefore to remove their normative halo. It can then be put back only as Thomas Hobbes did, via the implausible and ad hoc doctrine of "might makes right."

The Greeks envisaged morals as a system of rules, more like prescriptions for the successful practice of an art than like explicit and rigid rules of a game, or of metrical composition. It was no sign of linguistic poverty that one adjective, *kalos,* served the Greeks as the most general term of both aesthetic and moral commendation.

Nothing is said in the *Nicomachean Ethics* about law other than human law, and when this is spoken of, as in Book 5 on Justice, it is in the context of making the point that observing the law of the State is characteristic of one who possesses the virtue of justice. Greek moral philosophy is "practical," that is, concerned primarily with the question of what is the best way to live. One who has this knowledge and uses it is a virtuous person. And virtue, for Aristotle (who here as elsewhere is typical), is "a disposition concerned with choice, lying in a mean relative to us" (*Nicomachean Ethics,* 2.6.1106b36); the virtuous person follows a rule, but not a rigid or fixed one. Virtue is manifested in the exercise of judgment. In hard cases, one is counseled to imagine what the man of practical wisdom would do, and then to do it. No notion of obedience to law is involved. (Socrates' "I will obey the god, not you [Athenians] (Plato, *Apology* 29d) is no real exception. The obedience in question was to a particular specific command, viz., "Pursue your mission.")

All this was fundamentally changed by the triumph of Christianity, a sect of Judaism inheriting its divine-command moral theory. Morality henceforth meant obedience to the commandments of God (in reality, of the priests), with conformity to rituals and—for the first time—conformity in beliefs attaining equal importance with abstention from *mala in se*, the prohibitions of low morality. Indeed, even greater importance: you would be burned for heresy, merely hanged for murder. The high still trumped the low.

The true moral motive is inner: one does the right thing because it is the right thing. When morality is construed as commands, however, the motive for moral behavior becomes extrinsic, as it is in paradigmatic command/ obedience situations: one must obey in order to reap the rewards of fealty and to escape the punishment for defiance. In Kantian terms, the Greeks recognized morality as autonomous, whereas Judaism and Christianity made it heteronomous.

FURTHER CONSEQUENCES

Christianity, despite continuing political influence in Ireland, the United States, and elsewhere, is now intellectually defunct. But that does not mean things are back where they were before it took over. Two respects in which its influence is still strong are (i) the notion of moral law, and (ii) the notion of the pervasiveness of morality: in particular, that all sorts of actions, and even beliefs, are subject to moral appraisal.

The persistence of Christian moral theory is very evident, for example, in Kantianism, which finds the Moral Law Within to be of equal awesome-ness with the Starry Heavens Above. Kant emancipated himself from Pietism and from theology, but he continued to take it for granted that what every rational being, however untutored, recognizes within is a *law* that *commands* him or her to do this and abstain from that. Kant analogized this law to the laws of nature (another unfortunate metaphor) and postulated Reason as the legislator. There is truth behind this, namely, that the low morality is reasonable; that is, its observance is necessary for the successful prosecution of human ends. But Kant's way of putting it casts Reason in a dubious role as a commander. Moreover, if moral law is like (scientific) law of nature, it must be exceptionless; hence, the model encourages the delu-sion that to be moral is to be unswervingly obedient to rigid prescriptions. This comes out clearly in Kant's disastrous casuistry: since the moral law forbids lying, one must not lie even to save an innocent person's life.

These awkwardnesses in Kantian moral theory are consequences of the incompatibility that we have noted at its core, between the concepts of law (construed as command) and autonomy. It was to no avail that Kant made the law in question one that the moral agent prescribed to himself.

But it was not only Kant who carried over the Judeo-Christian assimila-tion of law to command into philosophical theory. No book of moral philosophy could be more resolutely secular than Jeremy Bentham's *Principles of Morals and Legislation*—which, nevertheless, famously begins with a paragraph describing the *subjection* of mankind to the *governance* of the two *sovereign masters*, pleasure and pain, to whose *throne* the standards of right and wrong are fastened. Thus, the notion of moral law infects the other main system of ethical theory too.

Indeed, so tenaciously (and unconsciously) do secular moral theorists hold to the command model, that—God and Reason having been rejected—*society* is recruited to fill the role of commander (thus obliterating the distinction between morality and the law of courtrooms); and when even this expedient proves inadequate (e.g., what does Society command about abortion?), moral prescriptions are construed as commands issued on speculation, as it were, by the individual utterer hoping that someone will hear and obey. Thus, Charles Stevenson analyzes "This is good [right]" as "I approve of this and I want you to do so as well," attributing imperial megalomania to the most diffident venturer of a moral judgment, as Sartrean existentialists also do. Things having come to this pass, perhaps it is time for the command theory to be retired.

AN ADVANTAGE OF THE COMMAND THEORY

But after all, even though grave theoretical difficulties can be discerned in the Hebrew project of religion's swallowing up morals, it was—and still is—a stupendous practical success. Why?

From the evolutionary standpoint, low morality is a set of dispositions constituting a survival technique for a certain kind of gregarious and omnivorous animal that makes its living by hunting and gathering foodstuffs in a (usually nomadic) band rarely exceeding forty in number. Much of it can be summed up as "Love thy neighbor"; however, the neighbor in question is one among only thirty-nine people. Everybody else is "Them," members of other groups, rivals whom it would be folly to deal with in the same considerate manner.

A numerical limit on the scope of morality presented a problem only when, about five hundred generations ago, agriculture was invented. The resultant civilization necessitated tremendously larger social units. The problem was, how to make communities of them? The solution was the invention of *government*: a development of the "natural" marauding party of the hunter-gatherer tribe into a permanent concentration of force under central direction, which would keep the intra-communal peace according to more or less formally promulgated *laws*.

To exert force effectively, the governmental body must be united and hierarchically organized; and this order cannot itself be maintained by force. Further, it needs to be perceived by the governed as legitimate, as something more than a band of ruffians. Both ends were encompassed through *high beliefs*, a sanctifying mythology, a State church.

As high beliefs are insulated from the experience of coping, there is no evolutionary sorting of the true from the false among them, and in fact virtually all of them are false. This does not mean, however, that it is a matter of chance which high beliefs are adopted in a community and which are rejected. Some high beliefs are *edifying*: they make believers feel better or more vigorous or enhance communal solidarity. Bands whose high beliefs are edifying will have an evolutionary advantage over those whose high beliefs are not.

Religious beliefs—imagination of superhuman powers—are high beliefs. Furthermore, any group sharing high beliefs forms a community, to that extent. High beliefs are social glue. They were already there when the need arose for them to bond hunter-gatherer bands into agricultural (and subsequently urban) societies with governments.

Thus, from the beginning of civilization down to the close of the eighteenth century, Church (high beliefs and high morality) and State (guarantor of intra-communal order, based on low morality) were inseparable allies.

The extraordinary power of survival exhibited by the Jews is, then, to be explained in large part by the strength of their attachment to a codified, definite, and all-encompassing set of high beliefs. The divine-commandment theory meant that in Jewish thought *all* morality was dependent on the will of Yahweh; consequently, severance from Him would mean the collapse of literally all that made social living feasible. "If X did not exist, everything would be permitted" is ridiculous for X = Zeus; but with "Yahweh" as the value of the variable, it is a truism, within the system.

This feature of Judaism carried over into Christianity and was a cause of its triumph, as well as of its persistence through periods when it had to stand on its own, unsupported by governmental authority and force.

HIGH BELIEFS IN JEOPARDY

The social arrangement we have been describing is one in which morality—the recognized norms of what is right for people to do—has two components: *low*, the precepts whose general observance is necessary for any viable community; and *high*, the rituals and taboos enjoined by the high beliefs the sharing of which defines the community. Deviations are kept in check by the government, which itself derives its authority from the high belief in its divine provenance. Although this state of things is an innovation, compared to the genetically programmed morality of the hunter-gatherer band, it is traditional enough to be regarded as "natural" for human societies based on agriculture. It is stable; in China and India it persisted without essential change for millennia. Presumably, its Western forms would have shown equivalent staying power had it not been for the Greeks and their invention of science. The scientific worldview is an integration and extension of *low* beliefs. Like religion, it is an enterprise requiring imagination; but unlike religion, the scientific imagination remains connected to the facts of experienced reality by the bond of logical inference—logic itself being a set of low beliefs. The picture of things revealed by science turns out to be self-sufficient; that is, nature can be understood in its own terms without supplementation by supernatural agencies. In the process some high beliefs are directly confuted; others succumb to Ockham's razor.

NOW

This epic struggle is far from being over; witness the current resurgence of fundamentalism. Science shows high beliefs to be irrational from the standpoint of epistemology. But it is also a finding of science that high beliefs are socially useful, perhaps indispensable, and that we tend instinctively to hang on to them at almost any cost. Nevertheless, the history of the past four centuries is one of inexorable scientific advance, while religion has more and more "veiled her sacred fires." Where does this leave morality?

Where it *should* leave us, from the logical point of view, is back at square one. That is, it should leave us with the low morality intact, minus the high morality, the rituals and taboos which are the actions that are the right things to do given (counterfactually) the validity of the high beliefs. Science, the true view of how things are, can only confirm, never call into question, the low morality, which comprises the necessary conditions of human communal living. (There is such a thing as moral knowledge; it consists of this fact, which is as independent of "cultural determination" as the Pythagorean theorem.) And perhaps the demise of various religious taboos is not altogether regrettable.

This is congruent to the aim of the more thoroughgoing *philosophes* such as Diderot and d'Holbach, which was to abolish all high beliefs and the high morality based on them. They supposed that the low morality, which could be said to be precepts of reason, would then be left to extend its gentle sway over all mankind. For reason is universal. Rational people, confronted with the same evidence, come to the same impersonal conclusions, both theoretical and practical. Kant's "Kingdom of Ends" would cover the earth. This vision continues to inspire philosophers down to John Rawls.

If the argument of this chapter is correct, one reason why this vision remains visionary is that decline in religious belief entails concomitant fading of sense of community. After family, the church is, to its believers, the institution generating the strongest sentiments of communion. Within the community, morality, by and large, can be counted on to take care of itself. But the boundaries of community are marked by high beliefs.

The historical function of a universal church such as original Christianity was to expand the emotional scope of community to the point that geography became almost irrelevant. It has failed, not on account of specific weaknesses that might be absent from some other system of high beliefs, but because the very idea of a system of high beliefs has fallen into disrepute. The great irony of the high belief is that it can do its (sometimes) beneficial work only as long as the fact that it *is* a high belief—a (mere) imaginative construction—escapes the notice of its believers.

To be specific: Expiration of high (Christian) morality would involve changes in attitude from disapproval to indifference or even approval of most of the following: divorce, birth control, extramarital sex, abortion, illegitimacy, homosexuality, nudity, pornography, blasphemy. Shifts of attitude toward these things have occurred on a broad scale, not just among the intelligentsia but all down the line, so that some are reflected in legislation. Save for divorce and birth control, the changes have taken place almost entirely in the last thirty years.

If low morality were to go, people would be indifferent, or even sympathetic, to promise-breaking, theft, assault, murder, destruction of property, perjury (and lying in general), gross incivility, draft dodging, contempt of authority of all kinds. Common experience and, no doubt, statistical studies bear out increases in all these behavioral categories in recent years. On the whole, however, and except for the last two, it does not seem that these activities enjoy any significant amount of overt approval, although they are often *excused* when perpetrated by members of one community against those of another, if the first community is (really or allegedly) being oppressed by the second. And this is what is predicted by the hypothesis that low morality is built into everyone's genes but is felt as applying only within one's community, whereas high morality imposes obligations only on people who share a set of high beliefs.

SUMMARY

1. Animals, to cope, acquire information about their environment through their senses. Relying on this information, they act in such ways as

to maintain themselves and their species. Some actions are not instinctive or reflexive but involve choice. Evolutionary pressures tend to minimize propensities to make lethally wrong choices.

2. Gregarious animals cannot act habitually in ways incompatible with maintenance and efficacy of the group. Hence, they are biased to within-group choices that favor establishment and recognition of hierarchy, limitation of intra-group violence and rivalry, nurturing of the young, cooperation in obtaining and distributing food. These biases against disruptive activities we call *low morality*. They are basically the same for all gregarious animals, though they are more developed in species capable of more varied activities.

3. The low morality is most elaborated in human beings, because language makes possible kinds of cooperation (especially the giving and following of commands) that are beyond the capabilities of mute beasts. Language also makes possible imagination, hence stories, and beliefs not immediately associated with perception. We call these *high beliefs*. Among them are beliefs in the existence of powerful personlike agents responsible for uncontrollable and catastrophic natural occurrences. It is deemed possible and transcendently important to get on the good side of these agents by propitiatory sacrifices. Making these sacrifices may entail actions deemed indifferent by the low morality, or even contrary to it; but service to the gods takes precedence. Its obligatoriness is the high morality.

4. This was the situation generally in the ancient world: each community had a morality consisting of low and high components. The low component was the same for all. The high varied according to the local conception of the gods and their demands. No question of why it was right to follow the low morality ("Why be moral?") arose.

5. Among one people, the Jews, an innovation occurred. They recognized only one god, whom they conceptualized as their leader. Not only did He demand certain rituals, as all gods do, but the low morality was supposed to be obligatory because He had commanded it. Being moral, then, meant being obedient to the supernatural leader, and immorality was disobedience. He was further thought to reward and punish accordingly; at first, with earthly prosperity or misery; later, post-mortem. Taken up into Christianity, this command theory of morals became universal in the Western world. Utilitarian and Kantian efforts to rebuild morality on secular foundations fail because both still conceive of morality as a system of imperatives, which are unintelligible without commanders.

6. Hence, it was taken as obvious that religion and morals stand and fall together: morality must "expire" when or if religion does.

7. The rise of science affords a worldview dispensing with high beliefs, hence with religion. Christianity consequently suffers from terminal intellectual anemia. The question of whether morality must go down with it thus becomes pressing. It is patent that the specifically Christian (high) elements of morality are in fact disappearing. Low morality is not in so perilous a state, but it is endangered indirectly by the loosening of the bonds of community consequent on religious failure.

CHAPTER 24

༄

L'Envoi

If Hume had not preempted it I might have liked *A Treatise of Human Nature* as the title for this book. For as noted above, we both seek to give accounts of basic human beliefs and of why they are held. But Hume went at it inside out, working with only the meager tool kit of "ideas" allowed him by Descartes and Locke; whereas I could call on all the resources of present-day science, notably Darwinism, a treasury of established *fact* no longer disputed by anyone, save a majority of my countrymen.

THE DEMARCATION PROBLEM

As explained in the Introduction, the impetus to study the evolution of beliefs came from thoughts about the "Demarcation Problem," how to separate sense from nonsense, or the rational from the irrational, or serious candidates for being labeled "true" from frivolous surmises, or the epistemologically worthy from the unworthy, or however else one might characterize the dichotomy one desires to make. The high/low belief distinction supplies such a division on its surface—almost platitudinously; and with the bonus that it overlaps the true/false division very closely. At any rate I am unable to think of a *socially* high belief that might be true, or of a *socially* low belief that is false. (Of course *individually* high beliefs that are true abound, and there are a few false *individually* low beliefs.)

Socially high beliefs, though false, persist for the most part because they are edifying. Yet if they are to function this way, those who believe them must regard them as *true* in the ordinary way—as corresponding with reality. Thus in the societies where they are held, there must be mass deception, undetected or nearly so: there must be an invisible membrane separating the high from the low and protecting the high from being actually rubbed up against reality. Now generally one does well to be suspicious of hypotheses requiring the existence of mass deceptions. But in this case the customary deference to unanimous judgments is unwarranted. The evidence for the existence of the membrane in societies other than one's own is overwhelming, and the conviction that things must be uniquely different *here* will hardly survive a little reflection. This is not to deny, however, that there may sometimes be reason to think that the official high beliefs of some society may not be held with exactly the same kind or degree of conviction with which one believes that the sun will rise tomorrow—particularly in connection with an afterlife, as Hume and many others have noted.

MILESIAN SCIENCE

The theory of high and low beliefs makes possible a clear and concise description of the Milesian achievement as the first Grand Theory of Everything based on low beliefs. There were grand theories before Miletus—what we call mythologies. But the difference between them—all of them—and what Thales, Anaximander, and Anaximenes came up with is absolute. Mythologies are pluralistic: they postulate two or more kinds of reality, a nature and a supernature, people and gods, usually. They are externalistic: nature is shoved around by supernature. They are non-rational: their explanations do not even attempt to show how the events under consideration *had* to proceed as they did; they were results of supernatural free will. The Milesian science is the negation of all three of these characteristics. It is monistic, unified: there is only one kind of reality, the everyday world. It is naturalistic, immanentist: the forces that bring about change in nature are internal to that nature. It is rationalistic, necessitarian: there is an explanation (if only we can find it) of everything that shows why it *must* be as it is and do what it does (and being and doing are not ultimately to be distinguished).

These three postulates, which I have called the Milesian requirements, have been observed only in varying degrees by subsequent Grand Theories, as has been shown in the rest of this book. But science has approached them asymptotically, as it were: today within science there is no serious dissent from them as virtually defining the scientific enterprise.

It is natural to feel that big effects must have big causes, that so tremendous an event as the birth of science *could not* have occurred so abruptly at one time and place and as the work of only three men. But it did; the record is quite clear. And it never happened anywhere else: not in China, not in

India, not in Timbuktu. There is something to the Great Man theory of history after all.

SCIENCE AND PHILOSOPHY

There is a difference between doing science and commenting on what it is to do science. I have made that difference the basis of my distinction between the two disciplines. But since the same persons have often done both, not too much importance should be assigned to this essentially verbal decision, at any rate not for figures prior to recent times, when specialization has grown so intense.

Both scientists (e.g., Thales, Pythagoras) and philosophers (e.g., Parmenides, Spinoza), as well as scientist-philosophers (e.g., Aristotle, Descartes) have produced Grand Theories of Everything. How and to what extent the theories embody the Milesian requirements may serve as a convenient index of their acceptability, inasmuch as the True Theory of the Universe must respect all three: this at any rate appears to be a valid inference from the history of science.

THE IMPACT OF CHRISTIANITY

Christianity, imposed by force on the Greco-Roman world as the official Grand Theory of Everything, preserved from Platonism its soul/body dualism but consigned everything else in the Greek outlook to the rubbish heap. Suddenly the intellectual world was back where it was before Thales— but subservient to a mythology whose primary feature was the supremacy of an entity entirely foreign to Greek thought of any period, the Omnipotent Creator/Legislator. The world was held to be the creation out of nothing by this Being who could do absolutely anything that could be imagined, and who was constrained by nothing in His choices. There was therefore no longer anything acknowledged to be what it was of necessity, anything that "could not be otherwise." The whole world and everything in it was contingent on the inscrutable Will of the OCL.

Slowly and painfully, over thirteen centuries, the pagan Greek outlook revived, as the study of "second causes": God's purposes were inscrutable, but the physical means by which He brought them about might be investigated. This process, like reforestation after a volcanic eruption, might have proceeded to the point of full recovery had it not been for the passions aroused by religious Reformation and Counter-Reformation climaxing in the Galileo affair. But even that was resolved in the twentieth century when the Pope pardoned Galileo and ceased to oppose Darwinism as far as the human body is concerned.

It has been generally overlooked, however, that the doctrine of the contingency of the world stayed where it was, deemed indeed to be a purely logical matter: "logical possibility" as it came to be called, conceived to be a real kind of possibility in the area between contradictoriness—the absolute impossibility of the unimaginable—and the "physical" impossibility of "violating the laws of nature," another metaphor derived from the OCL, now applied to what the Greeks held "couldn't be otherwise" in the world *as it is*, which was to them the only kind of impossibility. This unclassical "logical possibility" and its offshoot, the notion of "possible worlds," pop up at crucial points in "modern" philosophy and explain its continuing divergence from science.

"LOGICAL IMPOSSIBILITY"

Philosophical Grand Theories may start out either outside-in or inside-out. The former begin with a survey of what we (perhaps naively) believe to be true of the world at large. They scrutinize it, rejecting the dubious bits; classify and systematize it; and come, finally, to give an explanation of the human mind and its capabilities in terms developed in that account. Thus Atomism tells us that the world is atoms in motion, and only at the end does it tackle the problem of how we can know this, given that our minds too must be soul atoms colliding, joining, and separating. On the other hand inside-outers hold that we must start from what we are directly and immediately aware of, consciousness and its contents, which is "in here," and infer from that what must be "out there." Among the ancients only Plato managed to produce a Grand Theory starting from inside; the Skeptics simply threw up their hands, declaring the gap from *seems* to *is* unbridgeable, or at any rate unbridged.

Descartes began inside, like the Pyrrhonists, but went further in (allegedly) doubting the very existence of anything beyond his own consciousness, because, he claimed, there *could* be—it was *logically possible* that there was—an Evil Demon deceiving him about even his belief that $2 + 3 = 5$. Descartes exorcised this bogey by the aid of an item that he found among the contents of his consciousness, the idea of "a substance infinite, eternal, immutable, independent, all-knowing, all-powerful, and by which I myself, and every other thing that exists, if any such there be, were created"—that is, the idea of the OCL.

Without going into the convoluted reasoning whereby Descartes "proved" the existence in reality of the OCL, it is evident that both the descent into solipsism and the rescue from it depend on acceptance of the notion of "logical possibility," which in turn could not exist without the prior idea of an OCL. No such argument could have been advanced, and no such demon could have been postulated, among the pagans ignorant of the OCL notion. Logical possibility, then, is not a basic logical concept but a holdover from medieval Christianity. Possible-world semantics and all the subtle controversies about how many possible worlds there are and where , are essentially continuations of the disputes of the schools.

The OCL is also the generator of the Problem of Induction, inasmuch as the crucial premise in Hume's argument to show that reason cannot validate inferences from the observed to the unobserved is that "we can at least conceive a change in the course of nature; which sufficiently proves, that such a change is not absolutely impossible."

Let there be no misunderstanding. The contention here is not that the phrase "logical possibility" denotes nothing; it is that what it designates, non-internally-contradictoriness, is not a species of possibility, any more than a teddy bear is a species of bear. True, the OCL may exist, as many people, including philosophers, believe, in which case logical possibility *is* a kind of possibility. (And He might endow teddy bears with life, so that they *would* be a species of bear.) After all, what "cannot be otherwise" depends ultimately on what the world *is*, and so possibility and necessity are empirical concepts, in a way. But philosophers of all metaphysical persuasions ought to get clear on the point.

As things stand, acceptance of logical possibility and possible worlds is a principal reason not only for continuing disputes about the subtleties mentioned above, but for continuing attention to, sometimes real skepticism about, the validity of induction, the existence of the "external world," personal identity, and so on, which bring philosophy as it is taught in institutions of higher learning into disrepute. For practicing scientists have no time for such trivia. Consequently, they scorn the advice of "philosophers of science" on how to conduct their affairs. This is a pity if, as this book has shown, in early times at least scientists and philosophers lived in close symbiosis.

Given, however, the flourishing state of the logical possibility industry among "professional philosophers," the view of the topic taken here is bound to be fiercely resisted, or worse, totally ignored. So let me make one more try to state it clearly and concisely.

1. It is incontrovertible *fact* that prior to Christianity there was universal agreement that possibility and internal consistency were separate notions, with overlapping but distinct extensions. Pegasus and the Chimaera were imaginable but impossible, period.

2. It is incontrovertible *fact* that the OCL, introduced with Christianity, was conceived as being able to bring about any imaginable state of affairs. The extensions of possibility and internal consistency thereby became identical.

3. It is incontrovertible *fact* that the possibility of a "change in the course of nature" is an indispensable premise of Hume's argument creating the Problem of Induction.

4. Therefore it is incontrovertibly a matter of *logical* (in the proper sense) *necessity* that rejection of the OCL entails the dissolution of the Problem of Induction—at any rate of the only argument ever advanced to establish it.

Further:

1. It is incontrovertible *fact* that doubt about the existence of the "external world" was an attitude not taken by anyone before Descartes.

2. It is incontrovertible *fact* that Descartes motivated his (alleged) global doubt solely by the hypothesis of the Evil Demon.

3. It is incontrovertible *fact* that Descartes modeled the power ascribed to the Evil Demon on the OCL, and that there could be no reason to even imagine the existence of any such being unless one antecedently conceded at least the possibility of the OCL. ("God can bring about everything that I clearly and distinctly recognize as possible." 4th Replies, Adams and Tannery 7, 219.)

4. Therefore it is incontrovertibly a matter of *logical necessity* that the notion of the Evil Demon presupposes that of the OCL and that, absent such presupposition, its introduction into any argument would be utterly inane.

5. Therefore it is a matter of both fact and logic that lack of evidence for, and plethora of evidence against, the existence of an Evil Demon shows the arbitrariness and futility of attempting to raise the question of the existence of the "external world."

INSTITUTIONS

The *institutional facts* (i-facts) to which John Searle directed attention in his famous paper "How to Derive 'Ought' from 'Is'" and subsequent publications involve second-order beliefs, that is, beliefs about or presupposing other beliefs. Institutions and their formations—which turn out to be more complex matters than can be contained within the Searlian formula "X counts as Y in context C"—are discussed in the chapter after that on Hobbes, whose *Leviathan* is a classic in the field. In particular, a sketch is presented of a Consent theory of State formation which does not make the unrealistic assumption, universal in other such accounts, that the participants in covenanting are equal in various respects, only that they prefer peaceful coping to fighting for supremacy.

Further attention is paid to the notion of Confidence, which plays such an important part in the Economy, and to the mysteries of fashion trends and the like unpredictables, many of which are subject to what I call the Python Principle: institutional viability that depends on the belief that others believe in it. A "real Problem of Induction" is discovered (but not solved) in this connection.

MORALS AND ETHICS

The chapter discussing these topics is placed after that on Hume because the account given is (I hope) in the spirit of *le bon David*. Like his, it takes the basis of morals to be feeling; though it goes beyond that to try to account for those feelings in the terms of *coping* at the center of the theory of high and low beliefs.

In most of this book, discussion centers on beliefs about what is the case. However, there are beliefs too about what *ought* to be the case, about how people ought to behave: a band's *mores*. Anthropologists find that mores are apt to differ from band to band, often strikingly. This gives rise to the notion of Relativism: what's right or good on this side of the mountain is different from what it is on the other side, and who's to judge? On looking closer, however, we find a core of beliefs universally held among people, and indeed all gregarious animals, about what restraints on conduct must be in force if living together is to be better on the whole than solitary existence; as well as further rules that differ from band to band, the mores observed in a given band being a melange of these two components.

These two kinds of rules, the universal and the local, bear an analogy to low and high beliefs, respectively, which, however, must be used with caution. The universal mores, like low beliefs, have been rubbed up against reality, here meaning communal prospering, and have been found to promote it. But so, on the whole, have the local mores. On further inspection, it is usually found to be the case that the local mores, like high beliefs, derive from religion and are favored by evolution accordingly as they are *edifying*. (Not all are; as religious values trump secular ones, practices such as cannibalism and human sacrifice, abhorrent to the low mores, may be nevertheless locally allowed.)

The question arises, Can a morality adequate to the needs of a community survive the dying out of high beliefs and consequent disappearance of their edifying function? The answer proposed is, Maybe.

CHAPTER 25

⌒∿⌒

Conclusion?

A few pages back I said that philosophy should concern itself with first and last things. So far this book has been about first things, or at most second or seventeenth. Now it is time to say something about last things. This will be a short chapter, as I cannot provide an upbeat ending, much as I would like to.

You remember that great *coup de cinema* in *2001: A Space Odyssey* where the caveman throws the bone whirling upward into the sky, replaced in the next scene by another elongated object—a spaceship. An exaggeration, but symbolically not extreme, of the astonishing speed with which humanity has come from a condition in which all people lived as only a few now do on the banks of the Orinoco and in the interior of New Guinea, to voyages into outer space. Ten or twelve thousand years. Cosmically infinitesimal.

But where did that caveman come from, and how long did it take for him to reach the point of being able to throw a femur into the air, or at another caveman? Still not *very* long—three or four million years.

In this book we have looked at the process in just a little more detail, lingering a little more at some crucial steps. I believe that overall it gives the impression of a certain inevitability. The development *had* to be among apes who walked upright and had opposable thumbs. They *couldn't* have been exclusively herbivorous; only by having to chase down their dinners could they acquire more intelligence than zebras and wildebeests. They *had* to be gregarious, because only in group living could language be invented. And then they *had* to use that invention to expand their imaginations to the point of thinking up (wrong) explanations for the things and events that terrified them.

Thus evolved, fitting neatly, even comfortably, into that condition of things we have called Eden, all of a sudden they emerged from that instinctive way of life they were so superbly adapted to, into cities and empires: Babylon. In two senses of the word, this was the beginning of history: records could be kept, and there were *events* to write about, involving *classes* (not just pecking orders), *priests* (not just shamans), *bureaucracies* (not just bossy people), *laws* (not just vendettas), *dynasties* (not just the Chief and his family for two or three generations), *soldiers* (not just everybody as the occasion arose), and *wars* (not just tribal skirmishes)—lots of them, almost continuously. In short, *civilization*.

Progress, as on the whole it was and is universally regarded. Who would want to be a hunter-gatherer again, lacking, as in Hobbes's thought-experiment of anarchy, "Industry,... Navigation,... commodious Building;... Instruments of moving, and removing such things as require much force;... Knowledge of the face of the Earth;... account of Time;... Arts;... Letters"? People, being the most adaptable of animals, began to learn by education—the *fast* way—how to cope with the new conditions; and given enough time, the abilities and attitudes needed might, by the *slow* way of organic evolution, have become instinctive, and Man a truly social animal.

But things did not work out that way. There had been civilization for only about ten millennia—and by no means everywhere—when there occurred the second major discontinuity, the *intellectual* one that we have been describing in this book: the invention of *science* in Miletus.

More progress? Again, universally regarded as such. Who would want to live as they did in Miletus: no electricity, no TV, no painless dentistry, no aircraft or even trains and cars, no penicillin, and so on and so on? These are products of technology, not strictly and directly of science; but they never could have come into being without science. And the science that brought them about, also, on its own, produced at least the beginning of an approximation to the True Theory of the Universe, surely the noblest achievement of the human spirit.

But wait. How will it all end?

Nature herself provides several scenarios. The ultimate increase of entropy to the maximum, when there is no more energy available to do work, and the burning out of the Sun's reserve of hydrogen, are absolutely inevitable but too far away for us to worry about. More asteroid collisions are highly probable and in a shorter time, but they might leave human survivors; and even, if Hollywood is to be believed, really advanced technology might cope with them.

What generate real worries are the consequences of the same human propensities and abilities that brought about science in the first place. This is the Final Paradox. Just in our own time it has become technologically feasible to exterminate all life, or at least all human life, on our planet:

quickly and (in a way) voluntarily, by nuclear war, or more slowly by irreversible global warming, turning Earth into the twin sister of Venus. And by Murphy's Law, it is hard to see how one or the other can fail to happen. *Just possibly* technology might come up in time with a non-polluting energy source. But aside from that, we need a miracle—which is something that science and philosophy have proved to be impossible.

So, it is very hard to resist the conclusion that it was not a good thing *for us* when Thales declared the All to be One, a Nature operating by its immanent energy, and comprehensible to Reason.

There may be optimists (I haven't met any) who will say, No matter; we are beginning to see that out of the gazillion galaxies, there must be a terabillion stars with planets, and of those a jillion with life on them, and at least thirty-seven googols where they have worked out the True Theory of the Universe.

True, no doubt; but it would be a comfort only if there were a View from Nowhere, which there isn't. All values are local. Life is good; but good *for us* necessarily is *here*.

Awaken to the clock radio playing Schubert's *Great C Major* symphony. No matter how teeming with life the rest of the universe may be, you can be certain that nowhere else but on *this* planet can this tremendous affirmation of life be heard. Nowhere else can anyone look at "The Luncheon of the Boating Party." Or read the line "And smale foules maken melodie." Or play the role of Mrs. Malaprop. Or be *your* true friend.

GLOSSARY

CONDITIONED REFLEX. Reflex behavior triggered by something learned to be correlated with the natural trigger. Examples: salivation upon hearing a bell; erection upon smelling Chanel No.5.

CONTINGENCY OF THE WORLD. Metaphysical doctrine according to which any fact about the world "could be otherwise" because its being as it is is dependent on the omnipotent God's will, which is in no way bound. This view, not held by any pagan Greek, was an importation into philosophy from Judaism via Christianity. See E. Gilson, *Spirit of Medieval Philosophy*, chapter 3, esp. p. 69 and note 1. He finds it in William of Auvergne and indeed in Clement, Epistle ad Corinth, 27, 4, (1st c. A.D.) God "has constituted all things by the word of His majesty and by that same word He could destroy them all."

COPING. Behaving in a way that does not lower the probability of surviving to the age of sexual maturity and of obtaining fertile mates. Examples: obtaining and consuming suitable food; fighting or fleeing, as appropriate, when attacked; courting.

GRAND THEORY OF EVERYTHING. A comprehensive worldview: mythological, scientific, or philosophical.

INSIDE-OUT PHILOSOPHY. A philosophical Grand Theory holding that the basic data must be the contents of consciousness: Plato's, Aristippus's, Descartes's, Hume's, Berkeley's.

INSTINCT. Tendency of an animal to engage the whole body in a certain activity typical of that animal at that stage of development. Modifiable or suppressible but with difficulty. Examples: suckling; migration; dam building; nest building; web spinning.

INTERNAL INDICATOR/REPRESENTATION. The information conveyed into the nervous system by the senses. It must be emphasized that notwithstanding the inevitable (so I fear) use of such picture words as "image," "representation," and "indication," and the word "information" itself, there is no such thing as "brain reading," something in the brain that is scanned by a "brain-reader" or "homunculus." The indicator is the sensor, and what it senses (or "reads" if you like) is the world. In animals above the reflex-only level, this sensing, if not inhibited, is a belief. This is not to say that in principle, and perhaps already in practice, by the use of CAT scans, MRIs, and such-like machinery, technicians may not be able to literally read the indicator. But if they do, they are in no way duplicating the experience that the sensing individual is *having* (not perceiving). For further discussion see Matson 1976, chapter 3.

KNOWLEDGE. True belief. This simple definition has been rejected ever since Plato, in favor of "*justified* true belief," which notoriously ran into difficulty when Gettier constructed a counterexample. But I hold, with Crispin Sartwell, that knowledge is the goal of inquiry regarding particular propositions, and justification has only the instrumental value of being the *criterion* that knowledge claims need to meet, not a necessary condition in itself. I would add that since languageless animals know nothing of justifications, adoption of the three-part definition would quite arbitrarily deprive them of eligibility to know anything, which I at least find counter-intuitive. If you believe it and it's true, you can't go any further, epistemologically.

MILESIAN or THALESIAN REQUIREMENTS. Three intuitions of desiderata for scientific worldviews, first discernible in the accounts of Thales, Anaximander, and Anaximenes. I have used various words to express them.

1. *Monism, Unity, Reductionism.* "The All is One," that is, at bottom there is only one kind of reality, in terms of which everything can be (ideally) explained.
2. *Naturalism, Immanence.* No basic distinction between what a thing *is* and what it *does*. Processes manifest the essential internal energies of things.
3. *Rationalism, Logos, Necessitarianism, Sufficient Reason.* There are no "brute" facts; everything is either self-explanatory or explainable in terms of other things; and explanation has as its ultimate aim the showing of how and why things "couldn't be otherwise."

PHILOSOPHY. "Talk about what it is to be reasonable" (Renford Bambrough). I think this is the core idea, though of course historically the word has meant many things, often including or even just being science. By this definition, philosophy need not be a purely a priori discipline.

PROPOSITION. A term of convenience to use when two or more sentences have the same meaning, that is, describe the same state of affairs (whether actual or not). I hold beliefs, not propositions, sentences, or utterances, to be the *primary* bearers of truth. For in evolutionary epistemology, beliefs (which uncontroversially may be true) come before language, both temporally and logically. Sentences, utterances, and propositions are means of conveying (describing) beliefs.

REFLEX. Unlearned, unmodifiable, stereotypical motion of a bodily part of an animal in response to a definite stimulus, the TRIGGER. Examples: iris expansion and contraction (trigger: variation in intensity of illumination); salivation (trigger: appearance of food); erection (triggers: various suggestions of a sexual nature).

RUB UP AGAINST REALITY. By this metaphor, used only of beliefs, I intend to signify a process similar to that of "verification" as used of propositions by positivists. A belief is rubbed up against reality when the circumstances envisaged in the belief occur, and the believer is, on account of his belief, either helped or hindered in coping with them. If helped (or rather, not hindered), that is evidence for the truth of the belief; if hindered, against. In either case, the belief thus tested has become a low belief.

Perhaps the use of the word "reality" here may seem question-begging. I mean by the word no more nor less than what is "out there," the world in its objectivity, what is independent of subjective opinion, what indeed makes opinions true (or false)—things-in-themselves if you like. The existence of a unique real world is not a supposition or presupposition or article of faith; it is simply the basic *fact*. To (purport to) bring it into question is to be a Cartesian skeptic, and I have shown (chapter 18) that Cartesian skepticism is derived from medieval Omnipotence theology, which is philosophically indefensible.

SCIENCE. The activity of using reasonable procedures to find out what things (including ourselves) are really like; and the product of that activity.

SIZING UP (A SITUATION). This somewhat colloquial expression for insight, grasping the import of, judging (or getting ready to judge), and so on, is, I think, well understood in its English usages. I do not intend to give it any "technical" sense, and any attempt to "analyze" it, at least by me, is bound to be inadequate in one way or another. But since I use the expression a lot, I suppose I ought to try to clarify my usage.

In my book *Sentience* I used the expression as a label for the things we can do that no digital computer could possibly do. Some of my examples were: get a joke; recognize a certain kind of misprint; grade an essay; explain a historical event; translate (certain kinds of texts); create a work of art (that would have counted as such in the good old days); construct an argument or scientific hypothesis; and philosophize. It was, I claimed, the nearest equivalent to the Greek verb *noein*, "grasp the significance of," as used by Homer and other archaic writers. If there were an English verb to go with the noun "insight," I would have used that. Sizing up a situation involves

1. Picking out, within the situation, certain features as relatively distinct;
2. Of these distinct features, recognizing some as more important than others and than those not distinguished (the background);
3. Of these important features, apperceiving the static or dynamic whole which they compose;
4. Relating this apperceived whole to one's interests;
5. Finally, sometimes, raising the question of what one is going to do about the situation. (*Sentience*, page 150.)

For further comments and discussion see *Sentience*, chapter 5.

TETHERING. This is a metaphor for relations between beliefs weaker than logical implication joining elements in a comprehensive theory and justifying the claim of these elements to have some connection to low beliefs. It is best illustrated by Anaximander's Grand Unified Theory, in which superficially disparate elements such as the cosmic hoops, the Big Bang, the drying up of the sea, and human evolution are brought together and connected to low beliefs (facts) such as the discovery of marine fossils on mountainsides, the anomalously protracted duration of human infancy, and the separative property of whirlpools. In more modern times the theory of phlogiston offers another example of tethering: though false, it is so closely connected to facts and, for the most part, explains them so satisfactorily, that it counts as real science. The modern theory of the Big Bang is likewise tethered to—not logically implied by—observations made by radio astronomers. Theories, on the other hand, invoking creative gods, final causes, "logical possibility," and the like, are untethered, free-floating in the heaven of pure imagination.

True tethering occurs only in the context of theories that observe the three Milesian requirements of monism, naturalism, and rationalism.

REFERENCES AND FURTHER READING

Materials enclosed in brackets { } in the text refer to this section.

Adam, Charles & Paul Tannery, eds.: *Oeuvres de Descartes*. Paris: J. Vrin, various dates.

Aristophanes: *The Clouds*. Trans. H. J. and P. E. Easterling. Cambridge, U.K.: Heffer, 1961.

Aristotle. *Complete Works*. The revised Oxford translation ed. Jonathan Barnes. 2 vols. Princeton University Press, 1984. Many other modern editions and translations exist, such as in Loeb Classical Library (bilingual), Harvard University Press.

Austin, John L. *Sense and Sensibilia*. Reconstructed by G. J. Warnock. Oxford: Clarendon Press, 1962.

———. *How to Do Things with Words*. Reconstructed by G. J. Warnock. Oxford: Clarendon Press, 1962.

———. *Philosophical Papers*. Ed. J. O Urmson & G. J. Warnock. Oxford: Clarendon Press, 1961.

Ayer, Sir Alfred Jules. *Language, Truth, and Logic*. 2nd revised ed. New York: Dover, 1946. (First publ. 1935.)

Bealer, George, *Quality and Concept* .Oxford: Clarendon Press, 1982.

Berkeley, George. *Three Dialogues between Hylas and Philonous*. First published 1713. Many editions exist.

Blom, Philipp, *A Wicked Company*, New York: Basic Books, 2010.

Braithwaite, R. B. "The Nature of Believing." *Proceedings of the Aristotelian Society* (new series) 33 (1929–33): 129–146.

Broughton, Janet, *Descartes's Method of Doubt*. Princeton University Press, 2002.

Burnet, John. *Early Greek Philosophy*. New York: Meridian, 1961.

Casertano, Giovanni. *Parmenide il metodo la scienza l'esperienza*. Naples: Loffredo, 1978.

Chalmers, David. *The Conscious Mind*. New York: Oxford University Press, 1996.

Chomsky, Noam. *Language and Mind*. New York: Harcourt, Brace and World, 1968.

Critchley, Simon. *The Book of Dead Philosophers*. New York: Random House, 2008.

Della Rocca, Michael. *Spinoza*. New York and London: Routledge, 2008.

Descartes, Rene.

Diels, Hermann, and Walther Kranz. *Die Fragmente der Vorsokratiker*. 6th ed. 3 vols. Berlin: Weidmann, 1961.

D/K: Abbreviation for preceding entry. Translation by the author.

Dreyfus, Hubert. *Being-in-the-World: A Commentary on Heidegger's Being and Time, Division I.* Cambridge (MA), MIT Press, 1990.

Einstein, Albert. *Physics and Reality* 1936.

Elster, Jon. "Belief, Bias and Ideology." In Hollis and Lukes.

Evans-Pritchard, E. *Witchcraft, Oracles and Magic among the Azande.* Oxford: Clarendon Press, 1937.

Farrington, Benjamin, *Greek Science.* Harmondsworth: Penguin, 1949.

Fogelin, Robert. *Philosophical Interpretations.* New York: Oxford University Press, 1991.

——— . *Walking the Tightrope of Reason.* New York: Oxford University Press, 2003.

Frankfort, Henri, et al. *Before Philosophy.* Harmondsworth: Pelican, 1949.

Fustel de Coulanges, Numa. *The Ancient City.* New York: Doubleday, 1864.

Geertz, Clifford. "The Growth of Culture and the Evolution of the Mind." In *The Interpretation of Cultures.* New York: Basic Books, 1975.

Gellner, Ernest. *Legitimation of Belief.* Cambridge, U.K.: Cambridge University Press, 1974.

——— . "Relativism and Universals." In B. Lloyd and J. Gay, eds. *Universals of Human Thought: Some African Evidence.* Cambridge, U.K.: Cambridge University Press, 1981. Repr. also in Hollis & Lukes.

Gettier, Edmund. "Is Justified True Belief Knowledge?" *Analysis* 1963, 121–123.

Gilson, Etienne. *The Spirit of Medieval Philosophy.* New York: Charles Scribner's, 1936.

Goody, J. *The Domestication of the Savage Mind.* Cambridge, U.K.: Cambridge University Press, 1977.

—— , and I. Watt. "The Consequences of Literacy." *Comparative Studies in Society and History* 5 (April 1963): 304–345.

Hadreas, Peter. *A Phenomenology of Love and Hate.* Aldershot: Ashgate, 2007.

Hahn, Robert. *Archaeology and the Origins of Philosophy.* New York: SUNY Press, 2010.

Hampshire, S. *Thought and Action.* London: Chatto and Windus, 1959.

Heath, Sir Thomas. *A History of Greek Mathematics.* 2 vols. Mineola: Dover, 1981. (First publ. 1921.)

Heidegger, Martin. *Being and Time.* Transl. Joan Stambaugh. New York: SUNY Press, 1996. (See also Dreyfus.)

Heil, John. *From an Ontological Point of View.* Oxford: Clarendon Press, 2003.

Hesse, Mary. *The Structure of Scientific Inference.* London: Macmillan, 1974.

Hobbes, Thomas. *English Works.* ed. Sir Wm. Molesworth.

Hollis, Martin. "The Social Description of Reality." In Hollis and Lukes.

—— , and Steven Lukes, eds. *Rationality and Relativism.* Oxford: Blackwell, and Cambridge, MA: MIT Press, 1982.

Holmes, Richard, *The Age of Wonder.* New York: Random House, 2008.

Horton, Robert. "Tradition and Modernity Revisited." In Hollis and Lukes.

Hume, David. *A Treatise of Human Nature.* First publ. 1739–40. Many editions exist, e.g. Oxford: Clarendon Press, 1896.

——— . *Enquiry Concerning Human Understanding.* First published 1748. Many editions exist, e.g. Oxford: Clarendon Press, 1896.

——— . *Dialogues Concerning Natural Religion.* First publ. 1779. Many editions exist.

Kahn, Charles H. *Pythagoras and the Pythagoreans.* Indianapolis: Hackett, 2001.

Leite, Adam. "How to Take Skepticism Seriously." *Philosophical Studies* 148, no. 1 (2010): 39–60.

Locke, John. *Essay Concerning Human Understanding.* First publ. 1689. Many editions exist.

Lucretius. *De Rerum Natura.* 1st century B.C. Many editions and translations exist.

Maeterlinck, Maurice. *The Life of the Bee.* Trans. Alfred Sutro. New York: Dodd, Mead, 1912.

Matson, Wallace. "The Naturalism of Anaximander." *Review of Metaphysics* 6 (1953): 387–395.

——— . *The Existence of God.* Ithaca: Cornell University Press, 1965.

——— . *Sentience.* Berkeley: University of California Press, 1976.

——— . "Parmenides Unbound." 1980. In Matson, *Uncorrected Papers.*

————. "The Zeno of Plato and Tannery Vindicated." In *La Parola del Passato* (Naples) 1988. Revision, "Zeno Moves!" In Matson. *Uncorrected Papers.*

————. *A New History of Philosophy.* 2nd ed. New York: Harcourt, 2000.

————. *Uncorrected Papers.* Amherst, N.Y.: Humanity Books, 2006.

Monod, Jacques. *Chance and Necessity.* New York: Knopf, 1971.

Mounce, H. "Understanding a Primitive Society." *Philosophy* 48 (1973): 347–362.

Mourelatos, Alexander P. D., ed. *The Pre-Socratics.* New York: Doubleday 1974.

————. *The Route of Parmenides.* Rev. and expanded ed. Las Vegas: Parmenides Publishing Co., 2008.

Nadler, Steven. *Spinoza: A Life.* Cambridge, U.K.: Cambridge University Press, 1999.

Nagel, Thomas. *The View from Nowhere.* New York: Oxford University Press, 1986.

Nietzsche, Friedrich. *The Gay Science.* Trans. Walter Kaufmann. New York: Random House, 1974.

————. *Beyond Good and Evil.* Trans. Walter Kaufmann. New York: Random House, 1966.

————. *The Birth of Tragedy* and *The Genealogy of Morals.* Trans. Francis Golffing. New York: Doubleday, 1954.

Plato. *Collected Dialogues.* Ed. Edith Hamilton and Huntington Cairns. New York: Random House, 1961. Many other editions and translations exist.

Proust, Marcel. *Swann's Way.* Transl. C. K. Scott Moncrieff. Random House 1934.

Rawls, John. *A Theory of Justice.* Cambridge, Mass.: Harvard University Press, 1971.

Riedwig, Christoph, *Pythagoras, His Life, Teaching, and Influence.* Transl. Steven Rendall. Ithaca: Cornell University Press, 2005.

Rousseau, Jean-Jacques. *The Social Contract.* First publ. 1762. Many editions exist.

Russell, Bertrand: *A History of Western Philosophy.* New York: Simon & Schuster, 1945.

Ryle, Gilbert. *The Concept of Mind.* London: Hutchinson's University Library, 1949.

Sartwell, Crispin. "Knowledge Is Merely True Belief." *American Philosophical Quarterly* 28 (1991): 157–165.

————. "Why Knowledge Is True Belief." *Journal of Philosophy* (April 1992): 167–180.

Schopenhauer, Arthur, *The World as Will and Idea.* First published 1818. Many editions exist, some with title *The World as Will and Representation.*

Searle, John R. *Speech Acts.* Cambridge, U.K.: Cambridge University Press, 1969.

————. *Making the Social World.* New York: Oxford University Press, 2010.

Sextus Empiricus. 2nd c. A.D. *Outlines of Pyrrhonism.* Many editions exist, notably in Loeb Classical Library.

Singh, Simon. *Big Bang: The Origin of the Universe.* New York: Harper, 2004.

Sperber, Don. "Apparently Irrational Beliefs." In Hollis and Lukes.

Spinoza, Baruch (or Benedict) .*Ethics Demonstrated in Geometrical Order* and *Treatise on the Improvement of the Intellect,* First publ. in *Opera Posthuma* 1677. Many editions and translations exist.

————: *Theologico-Political Treatise.* First publ. ca. 1670. Many editions and translations exist.

Stevenson, Charles. *Ethics and Language.* New Haven: Yale University Press, 1945.

Strawson, Sir P. F. *Individuals.* London: Methuen, 1959.

Stroll, Avrum. *Informal Philosophy.* New York: Rowman & Littlefield, 2009.

Stroud, Barry. *Hume.* London: Routledge, 1977.

Tannery, Paul. *Pour l'Histoire de la Science Hellène.* Paris: Alcan, 1887.

Tarán, Leonardo. *Parmenides.* Princeton: Princeton University Press, 1971.

Taylor, Charles. "Rationality." In Hollis and Lukes.

Thomas Aquinas, St. *Summa Theologiae.*

Tsouna, Voula. *The Epistemology of the Cyrenaic School.* Cambridge, U.K.: Cambridge University Press, 1998.

Warren, Thomas B., and Wallace Matson: *The Warren-Matson Debate on the Existence of God.* Jonesboro, Ark.: National Christian Press, 1978.

Wilson, Edward O., and Stephen R. Kellert, eds. *The Biophilia Hypothesis.* Washington, D.C.: Island Press/Shearwater Books, 1993.

White, Andrew D. *History of the Warfare of Science with Theology in Christendom.* New York: D. Appleton, 1898.

Wittgenstein, Ludwig. *On Certainty.* Ed. G. E. M. Anscombe and G. H. von Wright. Trans. Denis Paul and G. E. M. Anscombe. Oxford: Basil Blackwell, 1969–1975.

———. *Philosophical Investigations.* Transl. G. E. M. Anscombe 1953.

———. "A Lecture on Religious Belief." In *Lectures and Conversations.* Berkeley: University of California Press, 1966.

INDEX